# Manufacturing, Automation Systems and CIM Factories

# Manufacturing, Automation Systems and CIM Factories

Edited by


## K. Asai

Osaka Institute of Technology
Asahiku
Osaka
Japan

and

## S. Takashima

Technical Research Institute
Japan Society for the Promotion of Machine Industry
Higashikurume City
Tokyo
Japan

Technical Editor

## P. R. Edwards

Lecturer in IT
ICL Institute of Information Technology
University of Nottingham
UK



**CHAPMAN & HALL**
London · Glasgow · New York · Tokyo · Melbourne · Madras

**Published by Chapman & Hall, 2–6 Boundary Row, London SE1 8HN**

Chapman & Hall, 2–6 Boundary Row, London SE1 8HN, UK

Blackie Academic & Professional, Wester Cleddens Road, Bishopbriggs, Glasgow G64 2NZ, UK

Chapman & Hall Inc., One Penn Plaza, 41st Floor, New York NY10119, USA

Chapman & Hall Japan, Thomson Publishing Japan, Hirakawacho Nemoto Building, 6F, 1-7-11 Hirakawa-cho, Chiyoda-ku, Tokyo 102, Japan

Chapman & Hall Australia, Thomas Nelson Australia, 102 Dodds Street, South Melbourne, Victoria 3205, Australia

Chapman & Hall India, R. Seshadri, 32 Second Main Road, CIT East, Madras 600 035, India



Co-published in 1994 by Ohmsha Ltd, 3–1 Kanda Nishiki-cho, Chiyoda-ku, Tokyo 101, Japan and Chapman & Hall, 2–6 Boundary Row, London SE1 8HN, UK

Typeset in 10/12 pt Palatino by Best-set Typesetter Ltd., Hong Kong
Printed in Great Britain by Clays Ltd, St. Ives plc, Bungay, Suffolk

ISBN 0 412 48230 4



A catalogue record for this book is available from the British Library

Library of Congress Cataloging-in-Publication data available

∞ Printed on permanent acid-free text paper, manufactured in accordance with ANSI/NISO Z 39.48-1992 and ANSI/NISO Z39.48-1984 (Permanence of Paper).

# Contents

*List of contributors* ix
*Preface* xi
*List of abbreviations* xiii
*Introduction* xvi

**PART ONE The Current Status of Automation and CIM** **1**

**1 Introduction** **3**
1.1 Automated production technology and CIM development 3
*Kiyoji Asai*
1.2 Significance of factory automation and CIM 4
*Kiyoji Asai*
1.3 Systems engineering for the introduction of factory automation or CIM 8
*Kiyoji Asai*
1.4 NC machines and factory automation in Japan 12
*Masahiro Murakami*
1.5 Industrial robots in Japan 25
*Kanji Yonemoto*
Appendix 1.A Robot associations and societies throughout the world 52
Appendix 1.B International cooperation on scientific and technical information exchange on robotics 55

**2 Manufacturing systems technologies and CIM** **57**
***Satoru Takashima***
2.1 Progress towards CIM 58
2.2 Robotics and robot systems engineering 61
2.3 Japanese industrial robots 61

2.4 Numerical control engineering and factory automation 81
2.5 CIM systems and MAP/OSI in Japan 85
2.6 Manufacturing communication standards environment in Japan 90
References 95
Appendix 2.A Glossary of ISO definitions 96
Appendix 2.B Japanese Industrial Standards (JIS) 97

**3 Introducing manufacturing systems and CIM** **99**
***Okitoshi Tsunoda***
3.1 Competitiveness and productivity 99
3.2 The CIM concept 105
3.3 Implementation 110
3.4 Factory automation and CIM systems 118
3.5 Computer-aided quality control 128
3.6 CIM construction 133
3.7 Future implementation 135

**4 Safety in the CIM factory** **137**
***Noboru Sugimoto***
4.1 Safety technology innovation 137
4.2 Composition of the safety control system 138
4.3 Basic concept of safety confirmation technology 140
4.4 Operation of a safety-confirmation system in an automated factory line 145
4.5 Conclusions 152
References 152

**PART TWO Examples and Effects** **153**

**5 CIM systems in the machinery industry** **153**
***Shigeo Hotta***
5.1 Characteristic features of MAP 156
5.2 Practical realization of MAP in parts production plants 157
5.3 An experimental factory automation training system conforming with MAP 162
5.4 Conclusions 165

**6 FMS for the machine industry** **166**
***Akimitsu Nagae***
6.1 Application examples 167
6.2 Conclusions 171

**7 FMS in appliance manufacture** **173**
*Koji Saito*
7.1 The FMS concept in appliance manufacture 174
7.2 FMS evaluation measures 174
7.3 Progress towards FMS realization 175
7.4 Positioning and application in FMS 177
7.5 Conclusions 177

**8 Factory automation for control motor manufacture** **180**
*Toshio Kojima*
8.1 Factory layout 181
8.2 Automatic machining 184
8.3 Automatic assembly 186
8.4 Automatic inspection 187
8.5 Major technological developments 187
8.6 Economic efficiency of the control motor factory 192
8.7 Conclusions 192

**9 CAD/CAM in the automobile industry** **194**
*Toshiaki Takagi*
9.1 Product development systems 194
9.2 Development of CAD/CAM systems 195
9.3 Implementation of CAD/CAM 198
9.4 Future systems implementation 206
9.5 Conclusions 209

**10 CIM in computer assembly** **210**
*Hiroshi Watabe*
10.1 IBM and its global systems 210
10.2 The IBM Fujisawa plant 212
10.3 CIM and IBM 213
10.4 CIM strategy at the IBM Fujisawa plant 214
10.5 Fujisawa plant CIM 214
10.6 Benefits of CIM and the future 216

**11 Robots in mass-production automatic assembly** **218**
*Toshijiro Ohashi, Seii Miyakawa and Makoto Matsunaga*
11.1 Product design improvement 218
11.2 Line performance simulation 221
11.3 Automatic assembly for VCR mechanisms 224
11.4 Construction of automatic assembly stations 225
11.5 Improvement of line performance 229
11.6 Conclusions 230
References 231

**12 Robot spot-welding** **232**
*Junichi Kawamura*
12.1 Spot-welding robots 234
12.2 Arc-welding robots 236

**13 Robot painting systems** **241**
*Toshihiko Nishimura*
13.1 Current robotic painting 241
13.2 Features 242
13.3 Considerations before introducing painting robots 246
13.4 Future robot painting 249

*Index* 250

# Contributors

*Kiyoji Asai*
Osaka Institute of Technology
Osaka
Japan

*Phillip R. Edwards*
ICL Institute of Information Technology
University of Nottingham
UK

*Shigeo Hotta*
First Department of Research and Development
Toyoda Machine Works Ltd
Aichi-ken
Japan

*Junichi Kawamura*
Yasukawa Co. Ltd
Narashino
Japan

*Toshio Kojima*
Fanuc Ltd
Yamanashi-ken
Japan

*Makoto Matsunaga*
Hitachi Ltd
Yokohama
Japan

*Seii Miyakawa*
Hitachi Ltd
Yokohama
Japan

*Masahiro Murakami*
Japan Machine Tool Builders' Association
Tokyo
Japan

*Akimitsu Nagae*
Yamazaki Mazak Corporation
Aichi-ken
Japan

*Toshihiko Nishimura*
Kobe Steel Ltd
Kobe
Japan

*Toshijiro Ohashi*
Hitachi Ltd
Yokohama
Japan

*Koji Saito*
Matsushita FA Engineering K.K.
Osaka
Japan

*Noboru Sugimoto*
Mechanical Safety Research Division
Research Institute of Industrial Safety
Ministry of Labour
Tokyo
Japan

*Toshiaki Takagi*
Matsuda Motor Corporation
Hiroshima-ken
Japan

*Satoru Takashima*
Technical Research Institute
Japan Society for the Promotion of the Machine Industry
Tokyo
Japan

*Okitoshi Tsunoda*
Fujisawa Plant
IBM Japan
Fujisawa
Japan

*Hiroshi Watabe*
Fujisawa Plant
IBM Japan
Fujisawa
Japan

*Kanji Yonemoto*
Japan Industrial Robot Association
Tokyo
Japan

# Preface

This book describes the current state of automation of the Japanese manufacturing industry. When read from a Western-world point of view, it clearly identifies some of the cultural difference between Japanese and Western manufacturing industries, and identifies some of the reasons why Japanese manufactured goods have achieved a competitive edge.

One of the key issues is that Japanese manufacturing companies are more receptive to the philosophies and benefits of automation. Large-scale factory automation is by definition a high-cost activity which is highly disruptive to the roles of the existing workforce. It is also unreasonable to expect a short-term return on investment. These factors make such automation prohibitively expensive to smaller organizations unless they have assistance.

The Japanese Ministry of International Trade and Industry (MITI) and other government bodies actively encourage smaller manufacturing industries to invest in automation. Examples of this are the use of lease robots, special low-interest rate loans for investing in automation, and tax incentives for users of computer-controlled machines. The corresponding Western organizations such as the British Department of Trade and Industry (DTI) do not approach this level of support.

Japan also has a shortage of skilled labour which can be alleviated by the introduction of automation. In addition Japan has company unions, as opposed to the Western trade unions, which means that changing job specifications are viewed from a different perspective. These factors combine to minimize the workforce mistrust and opposition to automation which is prevalent in the West.

One of the great benefits of automation is the ability to manufacture small batches of a wide variety of products. This gives a competitive edge over the mass production of a small number of products used by traditional manufacturing companies.

When considering the automation of factories the approach of using the simplest equipment which can adequately perform a task is emphasized in this book. Thus the problems of inaccuracy, support and maintenance costs are kept to a minimum.

To summarize, the methods described in this book are applicable to a wide range of manufacturing companies worldwide. All of the examples quoted refer to actual implementations and thus there is concrete evidence of their viability. Hence, given the necessary funding, there is no reason why automation on the scale adopted by Japan cannot be adopted by all other industrialized nations.

Dr Phillip R. Edwards

# List of abbreviations

| | |
|---|---|
| ADCIMS | autonomous distributed/decentralized control integrated manufacturing system |
| AE | assemblability evaluation |
| AEM | assemblability evaluation method |
| AGV | automatic guided vehicle |
| AI | artificial intelligence |
| AI/AO | analogue input/analogue output |
| AMRF | automated manufacture research facilities |
| APT | automatically programmed tools |
| ARCLE | assembly robot control language |
| ARL | assembly robot language |
| ARTRA | advanced robot technology research association |
| AS/RS | automatic storage and retrieval system |
| ASAR | automatic storage and retrieval |
| ATM | automatic transfer machines |
| CAA | computer-aided assembly |
| CAD | computer-aided design |
| CAE | computer-aided engineering |
| CAI | computer-aided inspection |
| CAM | computer-aided manufacture |
| CAQC | computer-aided quality control |
| CAST | computer-aided stress test |
| CAT | computer-aided testing |
| CFM | continuous flow manufacturing |
| CG | computer graphics |
| CIA | computer-integrated assembly |
| CIT | computer-integrated testing |
| CNC | computer numerical control |
| CPU | central processing unit |
| CTS | conformance test specification |

| | |
|---|---|
| DASD | direct access storage device |
| DI/DO | digital input/digital output |
| DNC | direct numerical control |
| DPMM | discrete parts manufacturing model |
| DSS | decision support system |
| EC | engineering change |
| EDM | electric discharge machines |
| ESPRIT | European Strategic Programme for Research and Development in Information Technology |
| FA | factory automation |
| FGCS | fifth-generation computer system |
| FMC | flexible manufacturing cell |
| FMS | flexible manufacturing system |
| FTL | flexible transfer line |
| GPSS | general-purpose simulation system |
| HIM | human-integrated manufacturing |
| HMS | horizontal machining centre |
| I/O | input/output |
| ICAM | integrated computer-aided manufacturing |
| ICAR | International Conference on Advanced Robotics |
| INGCT | Institute for New Generation Computer Technology |
| IE | industrial engineering |
| IEC | International Electro-technical Commission |
| IFR | International Federation of Robotics |
| IIMM | integrated intelligent management manufacturing |
| IMM | intelligent management and manufacturing |
| IMS | intelligent manufacturing system |
| IR | industrial robot |
| IROFA | International Robotics and Factory Automation Centre |
| ISIR | International Symposium on Industrial Robots |
| ISO | International Standards Organization |
| IT | information technology |
| IUT | implementation under test |
| JCF | Joint Coordinating Forum |
| JETRO | Japanese External Trade Organization |
| JIRAS | Japanese Industrial Robot Association Standards |
| JIS | Japanese Industrial Standards |
| JIT | just in time |
| JMTBA | Japanese Machine Tool Builders' Association |
| LAN | local area network |
| LUNA | language for users' needs and aims |
| MAGVR | manufacturing automatically guided vehicle robot |
| MAP | manufacturing automation protocol |
| MC | machining centre |

| | |
|---|---|
| MELFA | MEL for factory robot language |
| MIR | manipulating industrial robot |
| MITI | Ministry of International Trade and Industry (Japan) |
| MMS | manufacturing message specification |
| MOF | Ministry of Finance (Japan) |
| MRP | material requirement planning |
| MTBF | mean time between failures |
| MTC | MAP test centre |
| MUM | methodology for unmanned manufacturing |
| MC | numerical control |
| NIST | National Institute of Standards and Technology (USA) |
| OA | office automation |
| OSA | open systems architecture |
| OSI | open systems interconnection |
| PASLA | programmable assembly robot language |
| PC | personal computer |
| PCB | printed circuit board |
| PLAW | programming language for welding |
| PCL | programmable logic controller |
| PSIM | personal sequential inference machine |
| PTP | point-to-point |
| QC | quality control |
| SCARA | selective compliance assembly robot arm |
| SCOL | symbolic code language (for robots) |
| SE | systems engineering |
| SERF | Sankyo easy robot formula |
| SI | system integration |
| SIMPOS | sequential inference machine programming and operating system |
| SME | shape memory effect |
| SME | Society of Mechanical Engineers (USA) |
| SMR | soft motion robot |
| SPEL | Suwa Seiko production equipment language |
| SQC | statistical quality control |
| ST | standard time |
| SUT | system under test |
| TQC | total quality control |
| VTR | video tape recorder |
| WAN | wide area network |
| WFMUG | World Federation of MAP/TOP Users Groups |
| WIP | work in process |

# Introduction

This book has been compiled to address the expanding fields of factory automation (FA) and the large-scale application of computers, computer-integrated manufacture (CIM). It examines these areas in terms of their significance, situation, perspective and basic system technologies with respect to the current situation in Japan.

The current automation technologies, current situation, and technology of computer-integrated manufacture is examined.

In order to provide a foundation for later chapters, each technology is examined with respect to the reasons for its usage and growth. Factory automation is examined as an automated production technology. Computer-integrated manufacture is examined as a computerized production technology. The systems engineering involved in the introduction of factory automation and CIM systems is outlined, and the growing use of numerical control (NC) machines and the development of robot management operations in Japan are described.

Chapter 2 deals with system networks, software databases, artificial intelligence, standardization and other recent technologies necessary for factory automation and CIM. Chapter 3 is a case study which illustrates IBM's successful introduction of CIM into the company. Chapter 4 identifies the risks accompanying the introduction of factory automation and CIM. Safety measures are suggested to counter these risks, and the effectiveness of the methods are evaluated. Chapters 5 through 13 describe actual examples. They show the introduction of factory automation and CIM into ten factories belonging to major Japanese companies. These experiences are related to the preceding four chapters.

Kiyoji Asai

# PART ONE

# The Current Status of Automation and CIM

# Introduction 1

## 1.1 AUTOMATED PRODUCTION TECHNOLOGY AND CIM DEVELOPMENT

*Kiyoji Asai*

In the past all production was manual. In order to improve quality and efficiency, human beings initially used tools and then machinery. More recently, automation and computers have been introduced into the production process; this introduction is significantly related to social and economic trends. The following paragraphs look at changes in production from social and economic viewpoints.

Factory production methods in Japan and other countries are varied and have developed since the end of the Second World War. Simple machine-operated production has been transformed into automated production. This has been accompanied by the introduction of computers, as typified by the recent use of intelligent information systems in production. Initially, mechanization, automation and the introduction of computers were introduced separately. They have now been integrated and used on a large scale in order to further improve efficiency. In the period of reconstruction after the Second World War, countries throughout the world strived for improved productivity. In the context of high-growth economy, each nation provided a large number of products. This was achieved via a narrow range of mass-produced products which were quickly absorbed by the marketplace.

The oil crisis of 1973 drove the world economy into a period of low growth. Markets were scaled down which led to excessive competition between manufacturers. Thus manufacturers were forced to supply market demand-driven products on time and at a low price. Products were varied in terms of specification, models, and dimensions to meet

*Manufacturing Automation Systems and CIM Factories*. Edited by K. Asai, S. Takashima and P.R. Edwards. Published in 1994 by Chapman & Hall, London. ISBN 0 412 48230 4

diversified market demands. In addition, there were frequent product changes to meet changing market needs.

The scaling down of the marketplace drastically reduced production quantities, increased product variety, and required the ability to fine-tune the output and variety of products in production. The result of this was that the previous narrow range, mass production methods were transformed into a flexible, wide range, small production pattern. At this time the flexible manufacturing system (FMS) was introduced, as an adaptable production system which met market demands. The Japanese defined factory automation (FA) as a general term for such flexible systems. This includes FMS, computer-aided design and computer-aided manufacture (CAD/CAM), the small-scale flexible manufacturing cell (FMC), simple robots, numerical control (NC) machine tools, and simple mechatronic machine tools.

Since the early 1980s, and despite the change from 'low' to 'stable' influencing social and economic situations, competition among manufacturers in the marketplace has become fiercer. In order to survive this competition, manufacturers have strived to develop new products and to enhance the efficiency of their organizations. Efficiency measures include energy conservation, reductions in stock surpluses and costs, improved quality, and the promotion of timely production. The objectives were to develop systems to meet the market demands and to supply products of higher quality at the right time and at lower prices. In this environment, computer-integrated manufacture (CIM) was put into use as a production system. It was closely linked to the market via sales and distribution departments in order to avoid wastage of personnel, money or goods. These systems had two aims: to optimize business income by integrating the entire company without compromising the flexibility of the individual departments; and to build a large-scale system using information network technology. System, computer and information network technologies have been developed to support the growth of CIM. Thus it can be said that CIM has provided a means of satisfying the requirement for manufacturers to be more competitive within the market.

Figure 1.1 shows the relationship between social and economic changes, factory automation, CIM, and other technological changes.

## 1.2 SIGNIFICANCE OF FACTORY AUTOMATION AND CIM

*Kiyoji Asai*

The term factory automation refers to flexible automation in the manufacturing industry. It includes computer-aided engineering (CAE) for product design, CAD/CAM for both product design and the administration and supply of parts to a production site, and various other

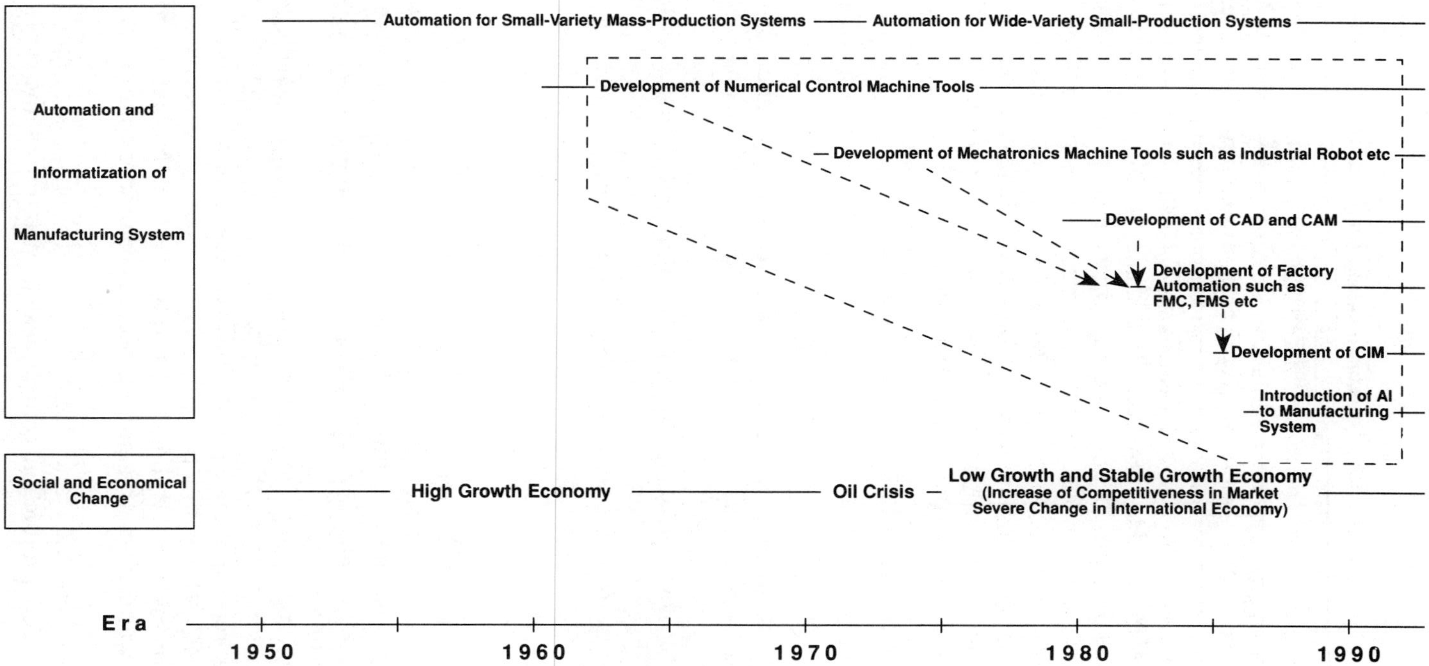


Figure 1.1 FA, CIM and social and economic change.

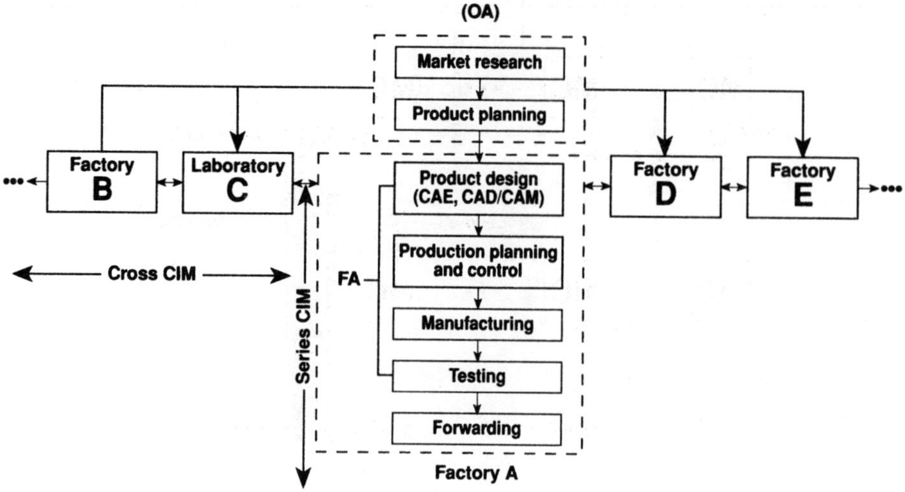


Figure 1.2 FA (including CAE, CAD/CAM) and CIM.

information systems. There are two types of CIM, 'CIM in series' and 'cross CIM'. 'CIM in series' describes the vertical application of computers and covers market research, product planning and design, production planning and control, testing and shipping. 'Cross CIM' describes the horizontal application of computers which builds bi-directional links between the office automation (OA) of headquarters and the many factories and business offices. The two types of CIM establish a company-wide production network with factory automation employed in actual production. These systems are shown in Figure 1.2.

Factory automation through the expansion of robot systems is achieved through FMS. The aim of FMS is to automate the entire factory and hence the term is often synonymous with factory automation. Figure 1.3 shows the hardware and software configurations, where each machining centre (MC) (a multi-functional machine tool capable of automatic tool change) performs a number of different machining processes.

Conveyor robots receive each workpiece from an automatic warehouse. They approach the appropriate MC, supply the piece for processing, retrieve it once processing is complete, and then proceed to the next MC. The entire procedure is repeated automatically. Once all MC processing is complete the robots return the pieces to the automatic warehouse. The entire process is computer-controlled.

CAD and CAM systems, or factory-wide production control systems, are added to the MCs and robots to provide software support. The computer-aided production control system determines the types and

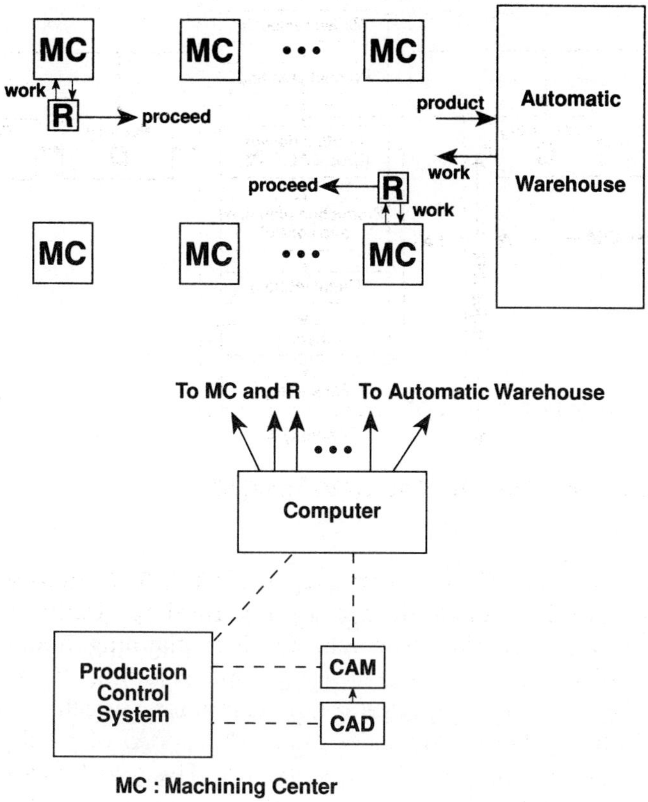


Figure 1.3 An example of a FA system.

numbers of products according to management strategies based upon market research, and plans production schedules. The hardware and software are still being developed with a view towards completely integrated systems. Although this particular example of factory automation is designed for assembly, it is anticipated that it will be applied to various other applications.

The relationships between the degree to which authority is centralized and independence is shown in Figure 1.4. The degree of centralized authority describes how densely authority is centred in the headquarters in terms of 'CIM in series' – the vertical linkage from headquarters to laboratory and factory. The degree of independence describes how independent each laboratory and factory can be in terms of 'cross CIM'.

At point A in Figure 1.4, the degree of centralized authority is low – authority is dispersed among factories and laboratories. Each laboratory and factory has a high degree of independence. Moving from point A

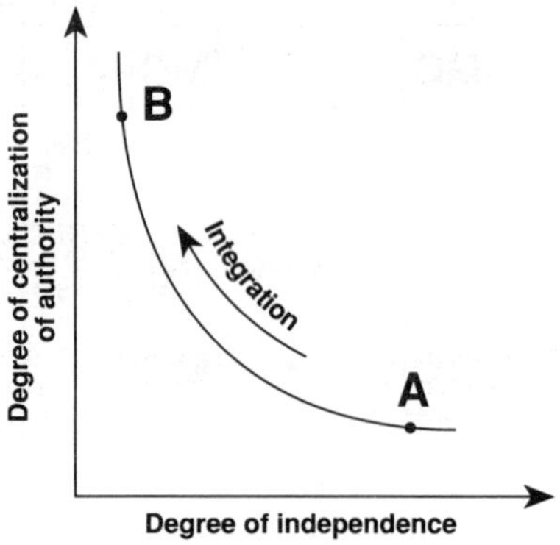


Figure 1.4 Explanation of integration.

towards point B, the level of integration advances and authority becomes more centralized at headquarters. The degree of independence of each factory and laboratory decreases. Around point B the entire company can be integrated in order to enhance its efficiency. This integration is achieved through an information network.

As company size increases, management and administration become increasingly more difficult. The traditional solution to this is to break up the company into manageable pieces by establishing corporate divisions or affiliated companies. Integration has been able to achieve success in the opposite direction, thanks to information technology.

## 1.3 SYSTEMS ENGINEERING FOR THE INTRODUCTION OF FACTORY AUTOMATION OR CIM

*Kiyoji Asai*

The development of the technology required for the practical introduction of factory automation or CIM into industry (i.e. systems engineering (SE)) lags far behind the recent rapid developments in equipment performance. To achieve maximum efficiency, SE must consider each piece of equipment as part of the whole rather than as an independent component. Many Japanese industries now appreciate the importance of SE towards obtaining an adequate return on the large investments in both factory and equipment associated with the introduction of factory automation or CIM. An increasing number of industries are employing the expertise of systems engineers in the introduction and utilization of factory automation and CIM.

The process of SE comprises a number of stages which are described in the following subsections. The first two stages are preparations before introduction. The actual introduction procedures, as shown

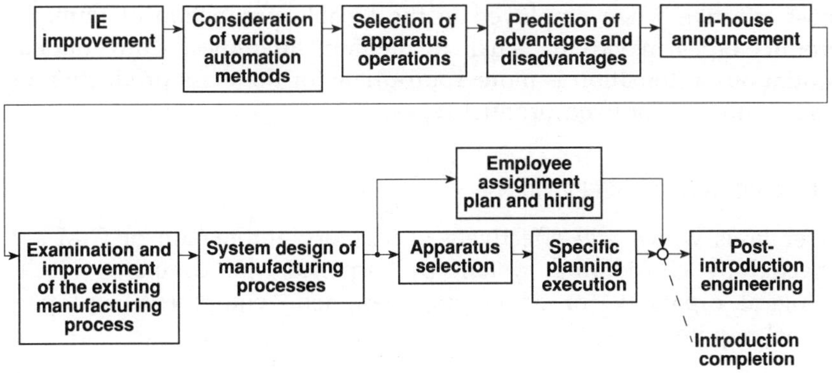


Figure 1.5 Procedures for the introduction of FA or CIM.

in Figure 1.5, then follow. The final stage is concerned with post-introduction engineering.

The stages outlined in the following subsections represent a set of general procedures for the introduction of factory automation or CIM in a factory.

### 1.3.1 EXAMINATION OF EXISTING EQUIPMENT

The purpose of this stage is to assess how far the existing systems can be improved. The optimum production efficiency of existing equipment must be carefully examined before introducing new automation or hardware. The possibility of making improvements to equipment layout, workforce assignments, production plans, process and quality control, worker development and performance, and worker morale must be considered. This type of production improvement requires no investment in either factory or equipment, constitutes an important preparatory step for future automation and acts as a basis for future plans.

### 1.3.2 REVIEW OF AVAILABLE AUTOMATION METHODS

When reviewing available automation methods, any means of mechanization, automation, information communications or instrumentation which may result in an improvement in productivity must be considered. The choice of the optimum equipment must be based upon production targets and economic feasibility. If the introduction of factory automation or CIM and related systems proves to be the most appropriate for a particular factory, then SE should be carried out.

Factory automation and CIM and their related systems are at their most effective when employed in the flexible production of multiple product types in varying small to medium quantities. Conventional continuous automation is more appropriate for the mass production of a small number of fixed product types.

### 1.3.3 OPERATION SELECTION

Operations which will effectively use factory automation or CIM are selected by considering system performance, the company's technological capability for developing peripheral equipment, and personnel requirements.

### 1.3.4 PREDICTION OF POSSIBLE ADVANTAGES AND DISADVANTAGES

Advantages may include increased productivity and improved working conditions. Disadvantages may involve difficulties in adapting working patterns to the new technology, workers' psychological alienation, and any physical dangers posed by factory automation or CIM.

The predicted advantages must specify initial target increases in productivity and improvements in working conditions. Adequate, comprehensive measures must be devised to counter any foreseeable disadvantages accompanying introduction.

### 1.3.5 IN-HOUSE ANNOUNCEMENT

Following the determination of operations and the clear identification and evaluation of the predictable advantages and disadvantages of factory automation or CIM introduction, the personnel involved should be informed. It is important to obtain the personnel's cooperation in examining the current conditions of selected operations, education and employee transfer. The personnel involved should be those engaged in the selected operations and other related procedures, their supervisors and representatives of the appropriate trade unions.

### 1.3.6 EXAMINATION AND IMPROVEMENT OF CURRENT MANUFACTURING PROCESSES

The current production flow, methods, and equipment should be examined in selected and related operations. This should be performed with the cooperation of supervisors and workers. The results of these examinations form a basis upon which improvements can be devised.

### 1.3.7 NEW SYSTEM DESIGN

A new production system design must integrate any possible improvements and process modifications, as determined by the examination of the current manufacturing situation. The total production system should be carefully considered and planned so as to include layout, worker assignments and production line balance (i.e. a balanced distribution of workload throughout the production line). Such planning should eliminate any stock or operating problems in a production system both before and after the introduction of factory automation or CIM. In addition, all the necessary equipment and safety devices must be designed for factory automation or CIM systems. A production plan needs to be devised and individual worker responsibilities determined.

### 1.3.8 EQUIPMENT SELECTION, STAFF PLANNING AND RECRUITMENT

The new system design is based upon general assumptions concerning equipment types and having the necessary staff. At this point, the equipment appropriate to the new system and staff number and skill requirements needs to be determined, and should be selected from the range on offer from manufacturers. Due to low return on investment and technical difficulties, only a limited range of standard equipment is available. The result of this may be that available models are over-specified for a particular operation or they may fail to meet requirements. In such cases it may be necessary to request modifications to a particular model. Alternatively a more expensive solution is to order a customized model which is designed for a specific application. Auxiliary devices and peripheral equipment can be designed and manufactured within the company.

The planning and recruitment of factory automation or CIM-related staff must commence as soon as possible in preparation for the initial operation. Staff requirements include operators and, if possible, maintenance and systems engineers. The systems engineer is responsible for solving problems associated with factory automation or CIM introduction, production system design, and software development. Further significant increases in productivity can be achieved by including an R&D engineer on the staff responsible for system improvement, new automation development, and the design and manufacture of auxiliary devices and peripheral equipment. Staff members who are recruited externally or from other sections must be trained prior to taking up their new assignments.

Shortcomings in either equipment selection or staff recruitment will necessitate modifications to the new system design.

### 1.3.9 CONCRETE PLANNING AND EXECUTION

This stage comprises:

(a) new system planning and design;
(b) the purchase of equipment;
(c) the preparation of auxiliary devices and peripheral equipment;
(d) the execution of the planned layout;
(e) the organization and assignment of employees;
(f) the testing operations and any adjustments.

Detailed planning between the previous stage and this stage will ensure a smooth transition from the old system to the new system.

### 1.3.10 POST-INTRODUCTION ENGINEERING

The new system, once put into operation in a factory, will initially require some further modifications and adjustments in order to achieve the target performance figures. Such aftercare is particularly important when adopting new technology. In addition, data and experience gained during the introduction of factory automation or CIM must be recorded and organized for future reference. This is especially true for subsequent introduction of factory automation or CIM into other production systems. These ongoing procedures are part of what is generally referred to as 'post-introduction engineering', which is a critically important task in its own right.

## 1.4 NC MACHINES AND FACTORY AUTOMATION IN JAPAN

*Masahiro Murakawi*

### 1.4.1 DEFINITION OF MACHINE TOOLS

Machine tools are often referred to as 'machine-making machines' or the 'mothers of machines'. This is because of their ability to accurately and cost-effectively produce the complicated precision parts needed for building all kinds of machines. Industrialized nations, which consider the machine tool industry to be the nucleus of their entire machine industry, make every effort to promote its growth.

Machine tools can be broadly defined as 'machines to process metals, wood or other materials into desired shapes by cutting, grinding, shearing, forging, rolling or other techniques'. The Japanese Industrial Standards (JIS) provide a narrower definition: 'machines which form primarily metallic materials into desired shapes by removing unnecessary parts by cutting, grinding or other techniques, excluding machines which are handheld during use or fixed with magnetic stands, etc.'. These statements show that in Japan, the term 'machine tool' usually refers only to machines which cut metals into desired shapes, and does not include forming machinery, etc.

### 1.4.2 CHARACTERISTICS AND CURRENT STATUS OF THE JAPANESE MACHINE TOOL INDUSTRY

Japan's machine tool industry is a cluster of small enterprises. According to the Japan Machine Tool Builders' Association (JMTBA), 82 of its 114 member companies have less than 300 workers involved exclusively in machine tool manufacture. In common with any other major manufacturing nation, Japan's machine tool industry is not very impressive in terms of production figures. It accounts for less than 2% of total machine production.

Traditionally, the machine tool industry has been oriented towards small-lot, diverse product production in response to the needs of user industries. The development of NC machine tools, though, has made possible the mass production of a narrower range of products. This is generating a contrasting trend towards mass production.

The structure of machine tool production began to change after the mid 1970s. More and more machine tool manufacturers, especially those producing NC turning machines and machining centres, developed overlapping product ranges. This made it increasingly difficult to classify companies by the type of machines they produced. When the economic environment improved during the 1970s, Japan's machine tool industry achieved significant growth fuelled by the development of superior NC machine tools. NC technology was introduced into Japan immediately after its original development in the USA during the 1950s. Since then, its development in Japan was strongly promoted. After 1975, the production of NC machine tools accelerated rapidly in Japan (see Figure 1.6 and Table 1.1). A number of factors contributed to Japan's machine tool boom. First and foremost, NC tools were able to satisfy diversity in users' needs. In addition, increased NC functions and reduced NC unit size, due to the introduction of microcomputers and quantity production capability, led to a lowering of NC machine tool prices.

Japanese machine tool production more than quadrupled in value between 1976 and 1985, from ¥ (yen) 229 billion to ¥1 051 billion. During the same period the production value of NC machine tools soared from ¥51 billion to ¥704 billion. This expanded the production value aspect of the NC machine tools share from 22% to 67%.

NC equipment became capable of combining the various lathe-turning functions which used to be provided by application-specific machines such as centre lathes, turret lathes, automatic lathes, etc. In

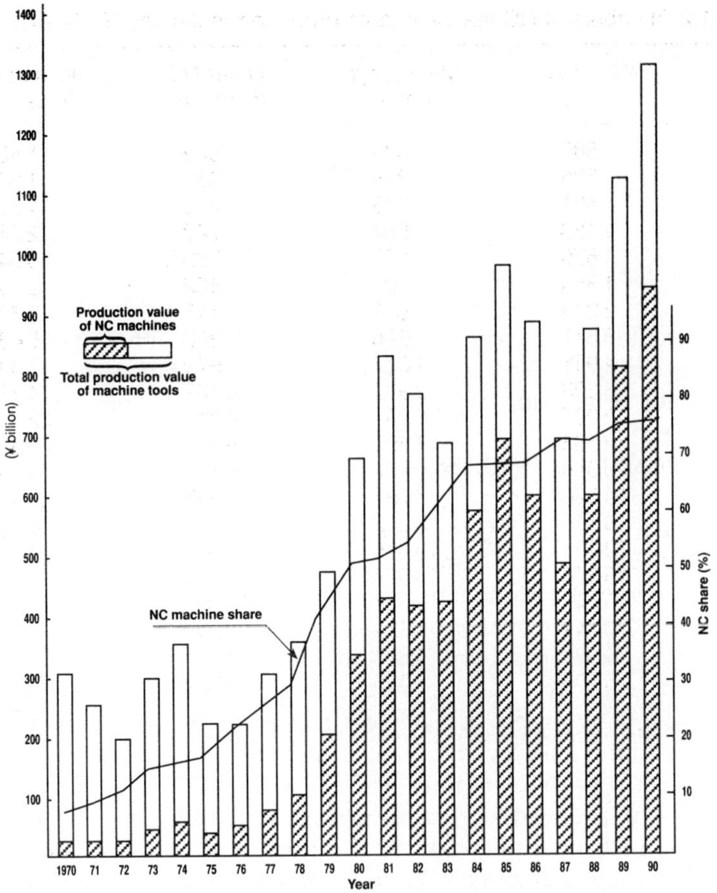


**Figure 1.6** NC share of total machine tool production in Japan, 1970–1990 (NC machine tools/total machine tools ×100%) Source: MITI *Production Statistics*.

**Table 1.1** Machine tool production shares in Japan (value in ¥ billions)

| | 1977 | | 1981 | | 1985 | | 1989 | | 1990 | |
|---|---|---|---|---|---|---|---|---|---|---|
| NC lathes | 42.7 | 13.7% | 161.3 | 19.0% | 221.0 | 21.0% | 251.7 | 22.1% | 288.8 | 22.2% |
| Machining centres | 23.1 | 7.4 | 165.4 | 19.4 | 267.3 | 25.5 | 286.2 | 25.1 | 335.2 | 25.7 |
| Other NC machines | 14.7 | 4.7 | 107.4 | 12.6 | 215.5 | 20.5 | 293.9 | 25.8 | 362.4 | 27.8 |
| Non-NC machines | 232.3 | 74.2 | 417.2 | 49.0 | 347.3 | 33.0 | 307.4 | 27.0 | 316.9 | 24.3 |
| Total | 312.8 | 100 | 851.3 | 100 | 1051.1 | 100 | 1139.2 | 100 | 1303.3 | 100 |

**Table 1.2** Numbers of NC machine tools produced in Japan, 1970–1990

| Year | NC lathes | Machining centres | Other NC machines | NC machines Total units |
|---|---|---|---|---|
| 1970 | 589 | 333 | 529 | 1451 |
| 1971 | 595 | 295 | 451 | 1379 |
| 1972 | 581 | 328 | 441 | 1350 |
| 1973 | 1459 | 564 | 742 | 2765 |
| 1974 | 1670 | 577 | 793 | 3040 |
| 1975 | 1355 | 401 | 432 | 2188 |
| 1976 | 2073 | 526 | 713 | 3312 |
| 1977 | 3677 | 926 | 833 | 5436 |
| 1978 | 4986 | 1377 | 979 | 7342 |
| 1979 | 8203 | 2927 | 3187 | 14317 |
| 1980 | 12007 | 5231 | 4814 | 22052 |
| 1981 | 12133 | 7394 | 6399 | 25926 |
| 1982 | 10344 | 6942 | 6852 | 24138 |
| 1983 | 10020 | 7791 | 8597 | 26408 |
| 1984 | 16555 | 10252 | 11229 | 38036 |
| 1985 | 19084 | 13345 | 12540 | 44969 |
| 1986 | 15976 | 10857 | 11943 | 38776 |
| 1987 | 15241 | 9027 | 11192 | 35460 |
| 1988 | 20942 | 11474 | 15234 | 47650 |
| 1989 | 24491 | 14828 | 18723 | 58042 |
| 1990 | 25406 | 15820 | 20739 | 61965 |

Source: MITI *Machinery Production Statistics*

addition a new type of machine, the turning centre, was developed with milling and other functionality. The combination of these developments began to supplant traditional non-NC turning machines.

Machining centres contributed to increased manufacturing efficiency due to their ability to perform different functions, such as drilling, boring and milling, automatically at a single workstation. Both NC lathes and machining centres became more popular in Japan, eventually accounting for nearly 50% of the total machine tool production value. Their growth in Europe and the USA has been moderate in comparison with Japan, but the market share of NC machine tools has also risen steadily in recent years (Table 1.2). In a similar way to Japan, technical progress and the increased use of NC machine tools are bringing about changes of unprecedented dimensions in the structure of machine tool industries throughout the world.

The machine tool industry has traditionally been strongly influenced by economic cycles. This industry, being a producer of capital goods, easily falls prey to recession and is one of the last beneficiaries of economic recovery. Machine tool orders remain depressed even after an economy begins to pick up. It usually takes six months to a year after the confirmation of a full-scale upturn before this industry begins to enjoy good business again. Machine tool manufacturers have learned to cope with this strong dependence upon economic cycles by adjusting back orders and labour forces, averaging production volumes, pursuing product specialization and coming up with unique products and offers.

A new worldwide problem is emerging for the industry which is of a more revolutionary nature and is completely independent of traditional economic cycles. This is a fundamental change in the structure of demand resulting from the increased use of NC turning machines and machining centres. The number of machine tool manufacturers has grown in response to the increased demand for these types of machine. Newcomers to the market have intensified competition and subsequently reduced profitability. At the same time the demand for the traditional product lines has been severely depressed. Faced with this situation, machine tool manufacturers throughout the world have been put under pressure during the 1980s to review their management strategies drastically.

### 1.4.3 MACHINE TOOL SUPPLY AND DEMAND IN JAPAN

Statistics released by the Japanese Ministry of International Trade and Industry (MITI) show that the value of machine tool production in Japan hit the ¥1000 billion mark for the first time in 1985. This slipped in 1986 and 1987 owing to an economic slump following the appreciation of the yen.

The decline in production of 1987 was especially marked. It fell to almost 70% of the 1985 level to leave the year's total value at a mere ¥690 billion. A ¥100 billion loss in export value was clearly responsible for this drop. This loss resulted from the import restrictions put into effect by the USA in January 1987, and from the disclosure of a COCOM violation incident in the May of the same year. These added to the general slump in the Japanese economy.

After mid 1987, machine tool orders began to pick up because key user industries, both in Japan and the rest of the world, invested heavily in machine tools in order to increase their competitiveness. There has been a boom in the Japanese machine tool industry in terms of orders, production, imports and exports since 1988.

#### (a) Orders

According to JMTBA statistics, machine tool orders in 1990 increased by 11.2% over the previous year to total ¥ 1412 billion, an all-time high.

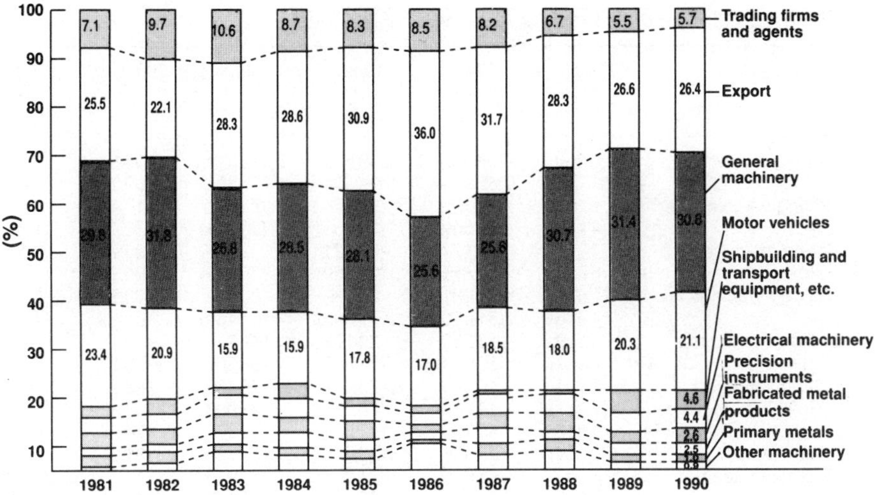


Figure 1.7 Machine tool orders by industry, 1981–1990.

Domestic orders accounted for ¥1039 billion, up by 11.5%. Overseas orders were ¥373 billion, up by 10.5%. These together indicate an expansionary trend led by domestic demand. The growth of NC machine tool orders was also steadily increasing by 12.5% in 1989 to ¥1174 billion, passing the ¥1000 billion mark for the first time. The percentage of machine tool orders which were for NC machines also rose from 82.2% in 1989 to 83.1% in 1990.

The general machinery and automobile industries deserve a special mention for their aggressive moves to expand production capacity, streamline equipment, and save on labour costs. Of the 12 key demand sectors, general machinery accounted for ¥436 billion (30.8% of total orders), followed by export orders with ¥373 billion (26.4%) and automobiles with ¥298 billion (21.1%). The export share of total orders has steadily declined by 9.6% over the five-year period from 36.0% in 1986, to 31.7% in 1987, 28.3% in 1988, 26.6% in 1989 and 26.4% in 1990.

The backlog in orders at the end of 1990 totalled ¥613 billion, up 16.8% from the end of 1989. This figure represents a 17% increase over the previous high of 1989, indicating that the machine tool manufacturers' production capacity could not keep up with increases in demand. The order backlog for NC machine tools also soared 14.3% over the last year, to total ¥451 billion (accounting for 73.6% of the total order backlog). This figure represents the equivalent of about 5.2

**Table 1.3** Machine tool orders by type of machine

| | 1987 (¥m) | 1988 (¥m) | 1989 (¥m) | 1990 (¥m) | 1990/1989 (%) | 1990 Share (%) |
|---|---|---|---|---|---|---|
| Turning machines | 200 560 | 288 388 | 359 625 | 390 017 | 108.5 | 27.6 |
| | (181 443)* | (262 944) | (334 544) | (366 906) | (109.7) | (26.0) |
| Drilling machines | 5 298 | 14 305 | 19 546 | 22 897 | 117.1 | 1.6 |
| | (4 155) | (12 649) | (17 399) | (20 901) | (120.1) | (1.5) |
| Boring machines | 14 122 | 25 177 | 33 821 | 32 691 | 96.7 | 2.3 |
| | (6 923) | (14 814) | (26 159) | (25 052) | (95.8) | (1.8) |
| Milling machines | 42 632 | 70 221 | 89 563 | 80 322 | 89.7 | 5.7 |
| | (31 446) | (51 400) | (67 993) | (57 469) | (84.5) | (4.1) |
| Grinding machines | 70 159 | 112 850 | 129 179 | 142 944 | 110.7 | 10.1 |
| | (24 544) | (50 534) | (71 865) | (86 366) | (120.2) | (6.1) |
| Gear cutting and | 12 774 | 16 515 | 20 647 | 26 460 | 128.2 | 1.9 |
| finishing machines | (7 211) | (8 321) | (14 021) | (22 033) | (157.1) | (1.6) |
| Special-purpose | 71 356 | 88 766 | 112 921 | 140 876 | 124.8 | 10.0 |
| machines | (21 611) | (25 490) | (39 255) | (51 728) | (131.8) | (3.7) |
| Machining centres | 175 646 | 245 676 | 321 217 | 377 917 | 117.7 | 26.8 |
| | (175 646) | (245 676) | (32 217) | (377 917) | (117.7) | (26.8) |
| Electrical discharge | 67 152 | 83 712 | 102 774 | 104 410 | 101.6 | 7.4 |
| machines | (62 692) | (79 416) | (93 475) | (99 264) | (106.2) | (7.0) |
| Other machines | 35 990 | 47 489 | 58 995 | 72 129 | 122.3 | 5.1 |
| | (19 330) | (27 698) | (35 757) | (44 714) | (125.0) | (3.2) |
| FMS (for shaft work) | — | — | 4 940 | 2 600 | 52.6 | 0.2 |
| | | | (4 940) | (2 600) | (52.6) | (0.2) |
| FMS (for prismatic | — | — | 16 454 | 18 860 | 114.6 | 1.3 |
| work) | | | (16 454) | (18 860) | (114.6) | (1.3) |
| Total | 695 698 | 993 099 | 1 269 682 | 1 412 123 | 111.2 | 100.0 |
| | (535 001) | (778 942) | (1 043 079) | (1 173 810) | (112.5) | (83.1) |

*Figures in brackets are for NC machine tools
Source: JMTBA

months of orders based upon average machine tool sales of ¥86.5 billion during the fourth quarter of 1990 (Figure 1.7 and Table 1.3).

### (b) Production

Machine tool production benefitted from strong private-sector equipment investment and increased by 14.4% in 1990 to an all time high of ¥1303.3 billion. Production of NC machine tools also soared by 18.6% to total ¥986.4 billion, another all-time high, and accounting for 75.7% of total machine tool production.

Production figures in 1990 totalled 196 000 units, an increase of 10.2% from the 178 000 units of 1989. NC machine tools accounted for 61 965 of these units, an increase of 6.3% from 1989.

Machining centres accounted for 15 820 units, value ¥335.2 billion (25.7% of total production value); NC lathes 25 406 units, value ¥288.8 billion (22.2%); grinding machines 13 055 units, value ¥156 billion

**Table 1.4** Machine tool production by type of machine

| | 1988 | | 1989 | | 1990 | |
|---|---|---|---|---|---|---|
| | Units | Value (¥m) | Units | Value (¥m) | Units | Value (¥m) |
| Turning machines | 28 257 (20 942)* | 227 035 (203 197) | 32 748 (24 491) | 281 321 (251 671) | 32 659 (25 406) | 317 402 (288 751) |
| Drilling machines | 31 731 (2 438) | 22 438 (15 468) | 37 705 (3 045) | 29 008 (20 190) | 38 670 (3 639) | 34 757 (24 831) |
| Boring machines | 1 223 (195) | 19 453 (9 644) | 1 447 (445) | 26 715 (18 942) | 1 501 (451) | 29 549 (2 207) |
| Milling machines | 6 959 (2 319) | 48 594 (30 176) | 8 612 (2 933) | 65 882 (42 001) | 8 492 (2 949) | 67 927 (43 447) |
| Grinding machines | 11 286 (1 447) | 105 742 (28 435) | 12 321 (1 929) | 139 157 (48 771) | 13 055 (2 428) | 156 016 (62 895) |
| Gear cutting and finishing machines | 1 001 | 16 623 | 1 080 | 21 672 | 1 172 | 24 835 |
| Special-purpose machines | 5 074 (1 640) | 103 569 (36 353) | 5 138 (1 640) | 131 735 (46 221) | 5 654 (2 087) | 160 154 (78 170) |
| Machining centres | 11 474 | 206 804 | 14 828 | 286 201 | 15 820 | 335 215 |
| Electric discharge machines | 6 140 (5 819) | 75 532 (71 947) | 6 923 (6 515) | 87 480 (83 425) | 6 939 (6 655) | 92 955 (89 751) |
| Other machine tools | 66 181 (1 376) | 55 278 (18 592) | 57 170 (2 216) | 70 037 (34 352) | 72 163 (2 530) | 84 472 (42 083) |
| Total | 169 326 (47 650) | 881 069 (620 615) | 177 972 (58 042) | 1 139 205 (831 773) | 196 125 (61 965) | 1 303 284 (986 351) |

*Figures in brackets are for NC machine tools
Source: MITI

(12.0%); and special-purpose machines 5654 units, value ¥160.2 billion (12.3%). The four types of machine represented 72.2% of total production value.

The buoyant production of recent years is attributable to a number of factors:

(a) industries are eager to invest in equipment in order to expand production capacity (this is especially illustrated by the general machinery and automobile industries);
(b) the eagerness of industries to make savings in labour costs and to combat labour shortages;
(c) the eagerness of industries to streamline production and to expand competitiveness (Table 1.4).

### (c) Import and export

Statistics released by the Japanese Ministry of Finance (MOF) show that the value of machine tool exports in 1990 totalled ¥455.8 billion. This 41.8% increase over the previous two years, which far exceeded growth, was attributed to a number of factors: a booming world

Table 1.5 Machine tool exports by type of machine

| | 1989 | | 1990 | | | |
|---|---|---|---|---|---|---|
| | Units | Value (¥m) | Units | Value (¥m) | 1990/1989 (%) | Share (%) |
| Physico-chemical process machines | 6718<br>(2950)* | 42278<br>(38998) | 7748<br>(2869) | 48637<br>(41805) | 115.0<br>(107.2) | 10.7<br>(9.2) |
| Machining centres | 5253<br>(5253) | 112070<br>(112070) | 4920<br>(4920) | 117483<br>(117483) | 104.8<br>(104.8) | 25.8<br>(25.8) |
| Special-purpose machines | 193<br>(133) | 18678<br>(13319) | 163<br>(108) | 18800<br>(15310) | 100.7<br>(115.0) | 4.1<br>(3.4) |
| Turning machines | 13148<br>(9266) | 129798<br>(120070) | 12786<br>(8674) | 137346<br>(129721) | 105.8<br>(103.8) | 30.1<br>(28.5) |
| Units | 103 | 356 | 42<br>(3) | 45<br>(20) | 12.6<br>(5.8) | —<br>(—) |
| Drilling machines | 7030<br>(1566) | 17423<br>(12984) | 7002<br>(1474) | 18297<br>(13872) | 105.0<br>(106.8) | 4.0<br>(3.0) |
| Boring–milling machines | 222<br>(107) | 5665<br>(4504) | 144<br>(58) | 3271<br>(2628) | 57.8<br>(58.3) | 0.7<br>(0.6) |
| Boring machines | 307<br>(73) | 4554<br>(2620) | 277<br>(69) | 7417<br>(4357) | 162.8<br>(166.3) | 1.6<br>(0.9) |
| Milling machines | 1719<br>(279) | 13667<br>(9334) | 1172<br>(275) | 13027<br>(8845) | 95.3<br>(94.8) | 2.9<br>(1.9) |
| Threading or tapping machines | 10466<br>(433) | 5423<br>(3010) | 15322<br>(484) | 7079<br>(4077) | 130.5<br>(135.4) | 1.6<br>(0.9) |
| Grinding or finishing machines | 13857<br>(890) | 55003<br>(26035) | 12789<br>(948) | 56507<br>(27685) | 102.7<br>(106.3) | 12.4<br>(6.1) |
| Other machine tools | 301041<br>(876) | 23676<br>(4959) | 359333<br>(263) | 27899<br>(6215) | 117.8<br>(125.4) | 6.1<br>(1.4) |
| Total | 360057<br>(21845) | 428591<br>(348250) | 421698<br>(20145) | 455809<br>(372016) | 106.4<br>(106.8) | 100.0<br>(81.6) |

*Figures in brackets are for NC machine tools
Source: MOF

demand for machine tools and shifting exports to east and southeast Asia.

The USA accounted for exports of ¥131.5 billion, down 9.2%; the EEC ¥134.1 billion, up 21.8%; and east and southeast Asia ¥136.1 billion, up 21.2%. These three destinations accounted for more than 88% of exports. NC machine tool exports increased 6.8% from the previous year to total ¥372 billion. Machining centres and NC lathes, both mainstay products, represented 54.2% of total exports (see Table 1.5 and Figure 1.8).

Meanwhile, imports made a healthy gain in 1990, increasing 35.9% from the previous year to total ¥68.6 billion. This increase appears to be attributable to the following factors: active domestic equipment investment; the value of the yen appreciated; American and European

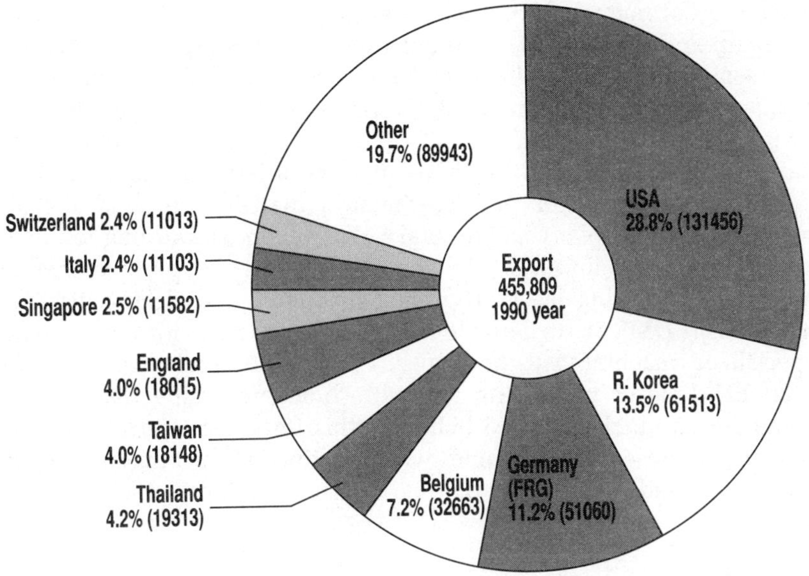


**Figure 1.8** Machine tool exports by destination, 1990 (values in millions of yen). Source: MOF.

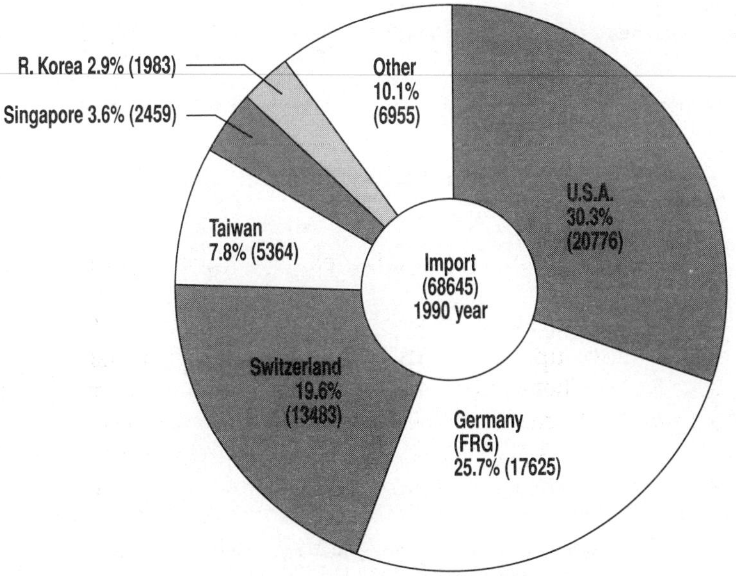


**Figure 1.9** Machine tool imports by origin, 1990 (values in millions of yen). Source: MOF.

machine tool manufacturers began to emphasize market development in Japan; and the domestic manufacturers' supply capacity decreased.

The main sources of imports were from the USA, accounting for ¥20.8 billion, up 7% from 1989; Germany ¥17.6 billion, up 94%; and Switzerland ¥13.5 billion, up 52% (Figure 1.9). These three countries have for years accounted for three-quarters of Japan's total imports.

The import value of physico-chemical machines increased 42.6% over the previous two years to total ¥17.6 billion, accounting for 25.9% of the total 1990 imports. The increased import of physico-chemical machines, including laser processing machines and electric discharge machines (EDM), is attributable to the increased need for precision and specialized machining technologies. Physico-chemical machines other than EDM, laser processing and ultrasonic processing machines account for as much as ¥10.3 billion of this import category. The functionality of these other machines was undefined at the time the statistics were compiled. The result of this is that machines which were not

**Table 1.6** Machine tool imports by type of machine

| | 1989 | | 1990 | | | |
|---|---|---|---|---|---|---|
| | Units | Value (¥m) | Units | Value (¥m) | 1990/1989 (%) | Share (%) |
| Physico-chemical process machines | 1 443<br>(303)* | 17 132<br>(5 712) | 1 530<br>(395) | 17 638<br>(6 700) | 103.0<br>(117.3) | 25.7<br>(9.8) |
| Machining centres | 103<br>(103) | 1 440<br>(1 440) | 164<br>(164) | 1 848<br>(1 848) | 128.3<br>(128.3) | 2.7<br>(2.7) |
| Special-purpose machines | 17 | 533 | 34 | 2 403 | 451.2 | 3.5 |
| Turning machines | 3 113<br>(74) | 4 620<br>(952) | 2 665<br>(83) | 5 580<br>(1 233) | 120.8<br>(129.5) | 8.1<br>(1.8) |
| Drilling machines | 32 865<br>(87) | 1 250<br>(360) | 24 865<br>(81) | 1 917<br>(386) | 153.4<br>(107.2) | 2.8<br>(0.6) |
| Boring–milling machines | 117<br>(31) | 831<br>(420) | 90<br>(14) | 769<br>(427) | 92.5<br>(101.7) | 1.1<br>(0.6) |
| Boring machines | 24<br>(3) | 276<br>(69) | 49<br>(5) | 892<br>(335) | 323.0<br>(483.6) | 1.3<br>(0.5) |
| Milling machines | 1 552<br>(56) | 1 354<br>(429) | 1 639<br>(54) | 1 996<br>(827) | 147.5<br>(192.8) | 2.9<br>(1.2) |
| Threading or tapping machines | 145 | 182 | 177 | 166 | 91.2 | 0.2 |
| Grinding or finishing machines | 62 288<br>(178) | 13 601<br>(6 265) | 48 413<br>(283) | 20 641<br>(10 549) | 151.8<br>(168.4) | 30.1<br>(15.4) |
| Other machine tools | 8 697<br>(47) | 9 276<br>(3 604) | 16 583<br>(73) | 14 796<br>(7 070) | 159.5<br>(196.2) | 21.6<br>(10.3) |
| Total | 110 364<br>(882) | 50 494<br>(19 253) | 96 209<br>(1 152) | 68 645<br>(29 376) | 135.9<br>(152.6) | 100.0<br>(42.9) |

*Figures in brackets are for NC machine tools
Source: MOF

actually machine tools were classified as such, due to the enforcement of the customs statistics classification of 1987 (under the Harmonized Commodity). Other than physico-chemical machines, the next most popular imports included grinding machines, value ¥20.6 billion; gear-cutting and finishing machines, value ¥9.8 billion; and lathes, value ¥5.6 billion. The import of machine tools totalled 1152 units, value ¥29.4 billion, up 52.3% from 1989 and indicating a steady upwards trend (Table 1.6).

### 1.4.4 SYSTEMS CONFIGURATION

The number of companies implementing FMS and factory automation systems has been on a steady upward curve thanks to the progress made in mechanical engineering and control technologies, and to improvements in reliability. Another contributing factor has been the reduction in staffing that can be achieved with these systems through unmanned operation at night. Most of the initial users of these systems, however, have been either the machine tool manufacturers themselves, or their affiliated companies manufacturing automobiles, industrial machinery and household electrical appliances.

MITI's 1987 'Report on the Seventh Survey of Machine Tools Installation' revealed that there were 259 flexible manufacturing systems (FMS) in operation at that time in Japan, of which 188 were metal-cutting FMS used in the machine manufacturing industries. According to a JMTBA survey, the use of flexible manufacturing cells (FMC) has been growing at a rate of 1000–1500 new systems per year.

An average FMS usually consists of five or six NC machine tools (principally machining centres) which account for about half of the total system cost. The other half of the cost comprises materials handling and transfer systems, storage, software and other components. Ongoing trends such as the renewal of obsolete NC machine tools in FMS and the transformation of combined speciality machines, including transfer lines into FMS and flexible transfer lines (FTL), should continue to boost system-based orders in the future (Tables 1.7 to 1.9).

**Table 1.7** Automated facilities in machine tool manufacturers in Japan: NC and non-NC machine tools (units)

| Years since installation | Total | <5 | 5–10 | >10 |
|---|---|---|---|---|
| NC machine | 2984 | 1536 | 1055 | 393 |
| Non-NC machine | 15139 | 1459 | 2430 | 11250 |
| Total | 18123 | 2995 | 3485 | 11643 |

Source: MITI *1987 Report of the Seventh Survey on Machine Tools Installation*

**Table 1.8** Automated facilities in machine tool manufacturers in Japan: FMC

| Base of FMC | Number of companies | Number of FMC systems | Average work-hours per month |
|---|---|---|---|
| Machining centre | 50 | 297 | 351 |
| NC lathe | 17 | 51 | 211 |
| Other NC machine | 7 | 15 | 267 |
| Total | 51 | 363 | 328 |

Source: JMTBA

**Table 1.9** Automated facilities in machine tool manufacturers in Japan: FMS

| | Number of companies | Number of systems | Number of machines |
|---|---|---|---|
| Shaft-work FMS | 8 | 12 | 44 |
| Prismatic-work FMS | 23 | 35 | 193 |
| Total | 25 | 47 | 237 |

Source: JMTBA

### 1.4.5 FACTORS INFLUENCING THE FUTURE OF THE MACHINE TOOL INDUSTRY

The emergence of new sources of demand and new types of demand within existing user industries will have a strong impact on future trends within the machine tool industry. There have been dramatic advances in machine technologies, such as in the aerospace industry. New technology-intensive industries, such as opto-electronics, have emerged. These factors are bringing about new demands for highly sophisticated machine tools. This trend towards increasing technology levels is also supported by other factors such as the needs of new industrial segments and the utilization of new materials. There is also a demand for software-inclusive total engineering in systems configurations such as CIM and factory automation in the manufacturing industries.

The future impact of these developments on the machine tool industry is difficult to predict quantitatively. However, taking into consideration the enormous potential demand which is expected to emerge from these and other new developments, it is clear that future demands may be the driving force of the machine tool industry.

Although the future of the machine tool industry will be largely directed by the user industries' requirements, it must also take the initiative in the development and introduction of technological innovations in anticipation of user needs. In particular, technological developments are likely to aim in three directions:

(a) the enhancement of technologies for machine tool construction and of the elements enabling the commercial development of machine tools capable of high power, high rigidity, high precision, high speed, high accuracy and high multiplicity;
(b) the incorporation of sophisticated electronic technology, e.g. the development of machine tools with a fully conversational mode as well as learning control;
(c) advances in computer software and networking technologies leading to the development of a new generation of machine tools. These will be capable of using the manufacturing automation protocol (MAP) in response to the increasing orientation towards CIM and factory automation.

### 1.4.6 CONCLUSIONS

There is an increasing requirement for manufacturing industries to achieve effective, diverse, small lot production, so as to meet diversified user needs. Factory automation attempts to meet these requirements using equipment centred on machine tools which are integrated via computer networks with the rest of the factory and business management.

The majority of today's machine tools are designed as the automation of individual machines. Products which are geared towards total systems configuration are extremely rare. The incompatibilities between factory automation equipment are a barrier to the building of factory automation systems. Factory automation-related equipment must be comprehensively standardized in order to promote systems configuration for CIM and factory automation. At the same time, machine tool manufacturers must consolidate systems to develop and supply machine tools which are tuned to the characteristics of the individual users.

## 1.5 INDUSTRIAL ROBOTS IN JAPAN

*Kanji Yonemoto*

### 1.5.1 GENERAL VIEW

Over the past 20 years, the use of industrial robots in Japan has produced a number of social and economic advantages. These include improvements in productivity, improvements in working environments, the prevention of industrial accidents, improved product quality, and early returns on capital investment.

These social and economic changes stem from the fact that industrial robots are functionally flexible and versatile. They have a freedom of movement comparable to that of the upper limbs (arms and hands) of human beings. This has enabled the automation of small batch production systems. The introduction of industrial robots has changed the production system from a man–machine system to a man–robot–machine system. This change has released personnel from dangerous and unpleasant working environments.

There have been recent efforts to improve the popularity of industrial robots in Japan. The following four policies were introduced in 1980:

(a) the establishment of a leasing system, and of the Japan Robot Leasing Company, which is designed to encourage the use of industrial robots by small and medium-sized enterprises;
(b) the provision of special financing from the Small Business Finance Corporation and the People's Finance Corporation to small and medium-sized enterprises for the introduction of industrial robots which are designed to ensure workers' safety;
(c) the establishment of a special depreciation system for high performance robots which are provided with computers;
(d) the application of loans and leasing programmes by local governments to help small enterprises to modernize their equipment by introducing industrial robots.

A further two policies were introduced in 1984:

(e) the establishment of the FMS leasing company using government finance and special interest rates (the Japan Development Bank provides special low interest rate loans to the Robot Leasing Company);
(f) the establishment of a tax system promoting the investment in advanced equipment, which uses electronic devices, for smaller enterprises (the so-called Mechatronics Investment Promotion Tax System), allowing special depreciation or the special deduction of corporate taxes.

Industrial robots are seen as the core technology of CIM and factory automation. This is reflected by the following additional policies which are being implemented for promoting research and development.

### (a) Research and development of extreme environment robotics

This large-scale national research and development project was implemented in 1983 by the Agency of Industrial Science and Technology of the Ministry of International Trade and Industry (MITI). It is scheduled to run for eight years and the total government funding will be approximately ¥20 billion. Its objectives are to develop robots for use in extreme environments where human activity is difficult or impossible. Three environments are targeted: nuclear power plants, the ocean, and disaster areas. Current development efforts are on atomic power related robots, robots to develop off-shore petroleum development, and robots for use in responding to natural disasters. In order to help achieve these objectives, the Advanced Robot Technology Research Association (ARTRA) was established in February 1984. The association today comprises 18 private enterprises and two public organizations.

### (b) Tax incentive programme

A tax incentive programme for the promotion of key technology research and development (High-Tech Incentive Programme) was established in 1985. This is a means of promoting key technology research and development in six specially designated fields, including high performance robots. Under this programme, effective between 1 April 1985 and 31 March 1994, 7% of the value of assets acquired for the use in the designated R&D areas is tax deductible.

### (c) Establishment of the International Robotics and Factory Automation Centre (IROFA)

This non-profit making organization conducts joint research into robotics and factory automation and serves as an information centre in charge of advanced technological information systems. IROFA's research programmes are in collaboration with industry, government and academia. A further objective is the implementation of these activities on an international scale. The Centre was established in June 1985 with the approval of the Minister for International Trade and Industry.

The practical utilization of industrial robots made outstanding progress in Japan throughout the 1970s and 1980s, with a total of 385 000 industrial robots in use by the end of 1990. A substantial increase in the popularity of robots is expected throughout and beyond the 1990s (see Figure 1.10).

The production of industrial robots in Japan amounted to ¥78.4 billion in 1980 and ¥544.3 billion in 1990 (Figure 1.12). The demand for industrial robots in the manufacturing industry in 1995 and 2000 is forecast to be ¥765 billion and ¥1330 billion respectively.

It is expected that there will be a substantial demand for industrial robots in the non-manufacturing industries. Such markets include nuclear energy, ocean development, civil engineering, and service industries such as medical services and transportation. The total demand for industrial robots in manufacturing and non-manufacturing industries is forecast as ¥900 billion in 1995 and ¥1640 billion in 2000 (see Figure 1.12 and Table 1.10).

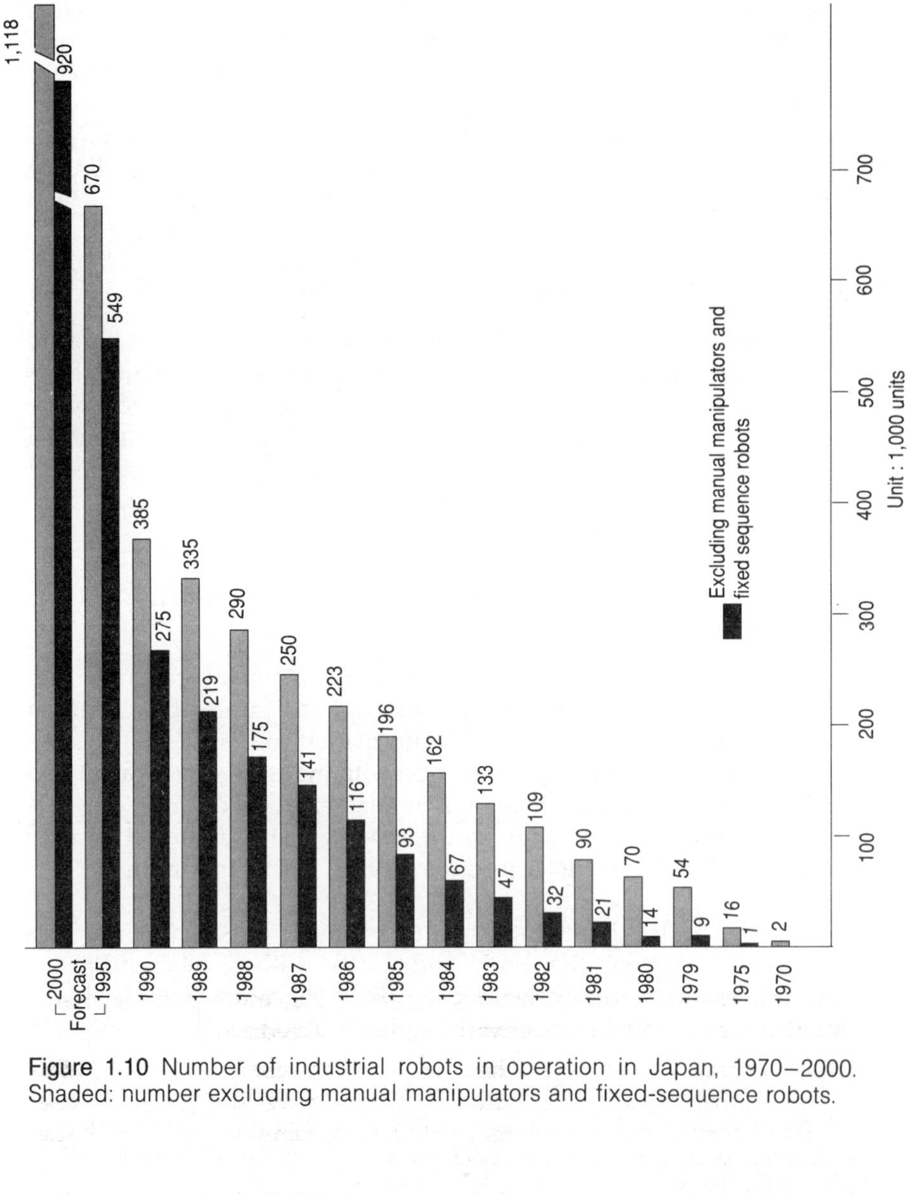


**Figure 1.10** Number of industrial robots in operation in Japan, 1970–2000. Shaded: number excluding manual manipulators and fixed-sequence robots.

Figure 1.13 shows the percentage production of industrial robots by classification.

### 1.5.2 DEFINITION OF AN INDUSTRIAL ROBOT

In the technical report ISO/TR8373:1988 of ISO/TC184 (Industrial Automation Systems)/SC2 (Industrial Robots in a Manufacturing Environment), a manipulating industrial robot is defined as follows:

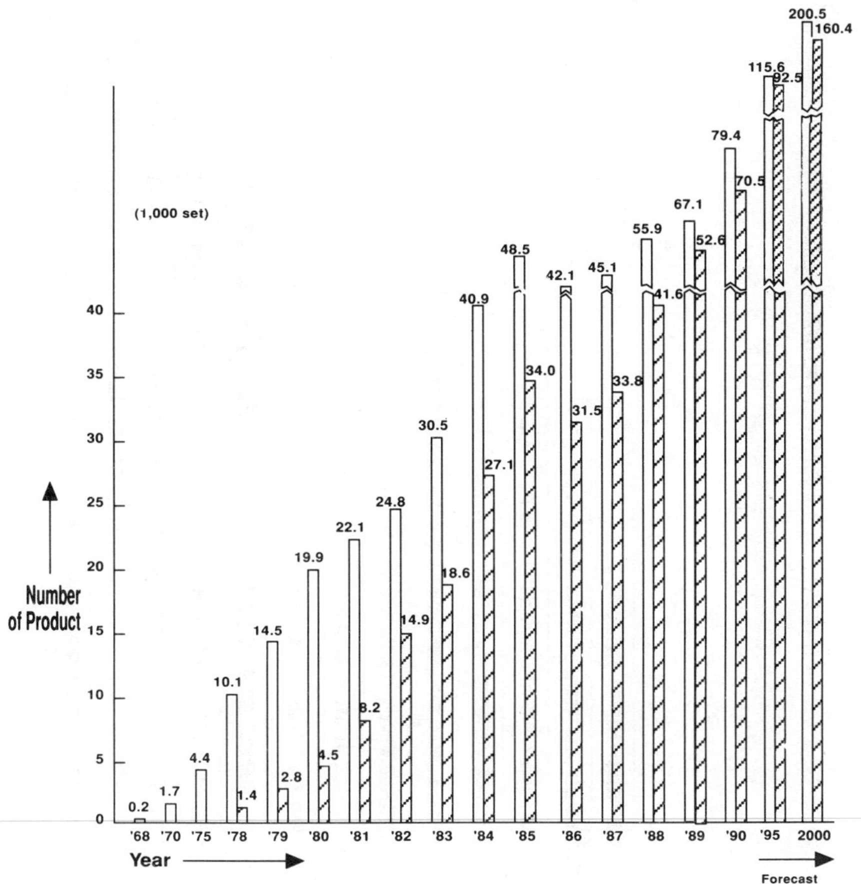


Figure 1.11 Production of industrial robots in Japan, 1968–2000. Shaded: number excluding manual manipulators and fixed-sequence robots.

> An automatically controlled, reprogrammable, multi-purpose manipulative machine with several degrees of freedom.

This definition is, however, limited to robots used in the manufacturing sectors. It does not include robots having locomotive or teleoperative functions, and robots used in non-manufacturing fields (i.e. not manipulating).

Industrial robots currently are devices which provide versatile and flexible movement functions similar to those of human limbs. In addition they may provide autonomous functions by means of sensing and recognition capabilities. It is anticipated that technological progress will include the development of devices with motion functions similar to those of creatures such as snakes and crabs. Therefore the classification and definitions of industrial robots will need to be modified as technology progresses.

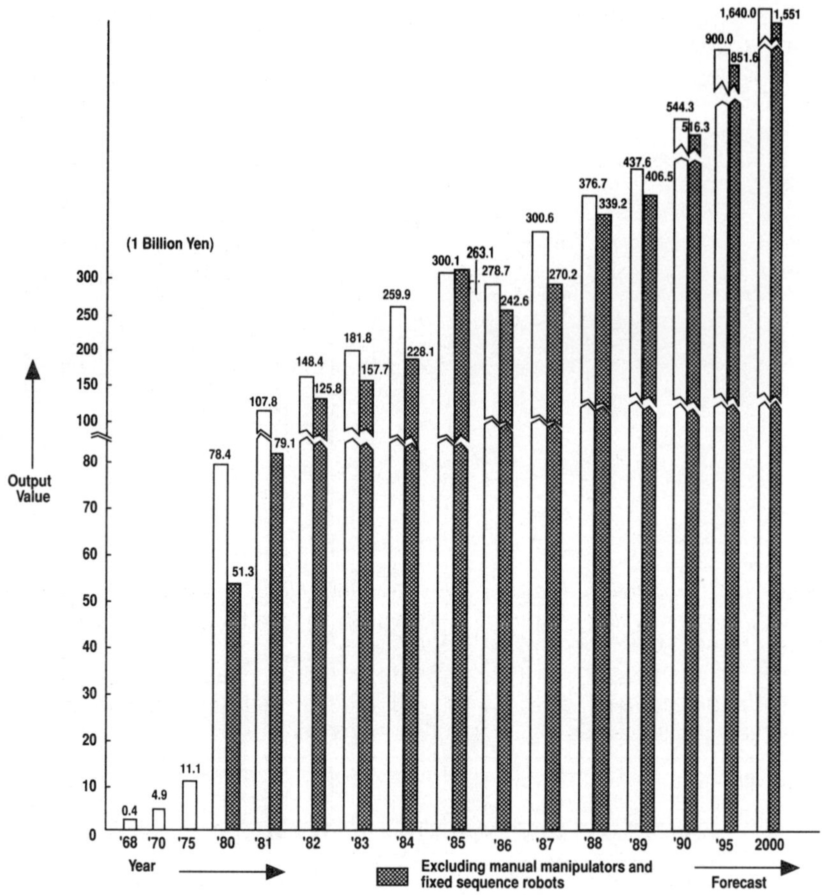


Figure 1.12 Value of annual robot production in Japan, 1968–2000.

The versatile, flexible motion functions of industrial robots have made the automation of smaller batch production possible. This has made a significant contribution to the improvements of productivity and product quality. The essential difference between industrial robots and other single-task automatic machines is that industrial robots possess flexible movement functions with a high degree of freedom in terms of both time and space. They can respond to changes in work-piece type, model changes, and design without the need for refitting. The result of this is that they are an effective tool for the automation of medium variety–medium volume and large variety–small volume production.

Table 1.11 shows the classification and definition of industrial robots according to Japanese Industrial Standards (JIS) B0134–1986, revised 1 May 1987.

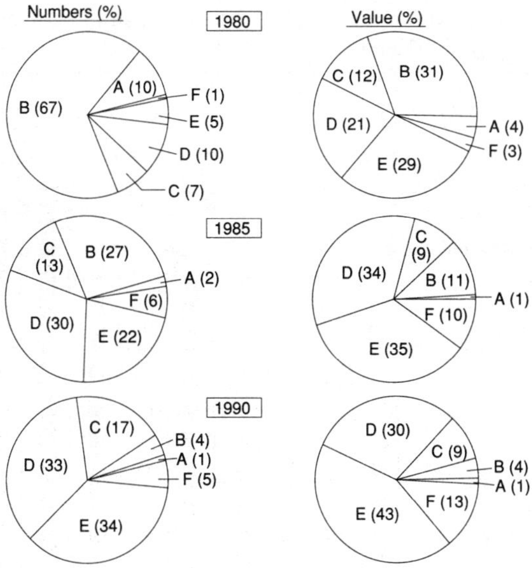


**Figure 1.13** Production of robots by type, 1980–1990. A, operating robots; B, fixed-sequence control robots; C, variable-sequence control robots; D, playback robots; E, CNC robots; F, intelligent robots.

**Table 1.10** Long-term forecast of demand for industrial robots in non-manufacturing industries in Japan (units: ¥100 million)

| Industries | 1995 | | | 2000 | | |
|---|---|---|---|---|---|---|
| | Min. | Max. | Average | Min. | Max. | Average |
| Agriculture, livestock | 39 | 78 | 59 | 167 | 203 | 185 |
| Forest | 16 | 76 | 46 | 28 | 84 | 56 |
| Ocean development, fishery | 122 | 187 | 154 | 369 | 415 | 392 |
| Civil engineering | 175 | 448 | 311 | 596 | 847 | 721 |
| Mining | 23 | 44 | 34 | 45 | 60 | 53 |
| Transportation, warehouse | 61 | 311 | 186 | 235 | 478 | 357 |
| Distribution, commerce | 8 | 89 | 48 | 20 | 100 | 60 |
| Gas | 2 | 26 | 14 | 16 | 39 | 28 |
| Water supply | 10 | 30 | 20 | 27 | 45 | 36 |
| Electric | 27 | 36 | 32 | 61 | 65 | 63 |
| Nuclear power | 103 | 136 | 119 | 148 | 157 | 152 |
| Communication | 4 | 11 | 8 | 14 | 21 | 17 |
| Space | 50 | 100 | 75 | 70 | 150 | 110 |
| Medical care, welfare | 32 | 44 | 38 | 117 | 125 | 121 |
| Garbage, cleaning, building maintenance | 13 | 29 | 21 | 81 | 97 | 89 |
| Fire fighting, antidisaster | 34 | 58 | 46 | 445 | 481 | 463 |
| Education | 17 | 27 | 22 | 42 | 49 | 46 |
| Research and development | 5 | 54 | 30 | 13 | 61 | 37 |
| Other services | 21 | 82 | 52 | 79 | 138 | 108 |
| Total | 763 | 1866 | 1315 | 2573 | 3616 | 3094 |

Source: JIRA survey (1985)

**Table 1.11** Classification and definitions of industrial robots (Japanese Industrial Standards JIS B 0134-1986)

| General Classification | Definition |
|---|---|
| 1. Operation robot | A robot with controllers and actuators for mobility and/or manipulation, remotely controlled by a human operator |
| 2. Sequence control robot | A robot that operates sequentially in compliance with preset information (sequence, conditions and positions) |
| 2.1 Fixed sequence robot | A sequence robot for which the preset information cannot be easily changed |
| 2.2 Variable sequence robot | A sequence robot for which the preset information can be easily changed |
| 3. Playback robot | A robot that can repeat an operation on the basis of instructions concerning sequence, condition, position and other information imparted by moving the robot under operation control |
| 4. Numerically controlled (NC) robot | A robot that can execute the commanded operation in compliance with information on e.g. sequence, conditions and position loaded numerically or by program, without being moved |
| 5. Intelligent robot | A robot that can determine its own actions through artificial intelligence |
| 5.1 Sensory-controlled robot | A robot whose operation is controlled by sensory information |
| 5.2 Adaptive-controlled robot | A robot having an adaptive control function (Note: Adaptive control signifies control in which the characteristics of the control system change in response to change in the environment in order to satisfy the prescribed conditions) |
| 5.3 Learning-controlled robot | A robot having a learning control function (Note: Learning control signifies control in which appropriate operation is obtained reflecting operating experience) |

### 1.5.3 BACKGROUND TO THE RAPID SPREAD OF INDUSTRIAL ROBOTS IN JAPAN

A number of factors have affected the diffusion of industrial robots throughout Japan (Figure 1.10 and Table 1.12). The development of industrial robots and their practical applications was initially driven by

**Table 1.12** Number of industrial robots in operation by countries (excluding manual manipulators and fixed-sequence robots)

| | 1990 end | 1989 end | 1985 end |
|---|---|---|---|
| Japan | 274 210 (59.5%) | 219 700 (57.0%) | 93 000 (54.7%) |
| Singapore | 1 600[a] | 1 459 | 200 |
| Australia | 1 490 | 1 350 | — |
| New Zealand | 70[a] | — | 31 |
| China–Taiwan | 1 293 | 965 | 227 |
| Asia, Oceania total | 278 663 (60.5%) | 223 474 (57.9%) | 93 458 (55.0%) |
| USA | 41 304 | 37 000 | 20 000 |
| Canada | — | — | 1 032[b] |
| America total | 41 304 (9.0%) | 37 000 (9.6%) | 21 032 (12.4%) |
| Belgium | 1 638 | 1 403 | 975 |
| Denmark | 489 | 402 | 164 |
| Finland | 825 | 674 | 257 |
| France | 8 551 | 7 063 | — |
| Germany | 28 240 | 22 395 | 8 800 |
| Italy | 12 500 | 10 000 | 4 000 |
| Netherlands | 1031 | 888 | 350 |
| Norway | 527 | 493 | 323 |
| Sweden | 3 791 | 3 463 | 2 046 |
| Switzerland | 1 525 | — | 290 |
| UK | 6 418 | 5 908 | 3 208 |
| Spain | 2 197 | 1 751 | 688 |
| Austria | 860 | 840 | 170 |
| Western Europe total | 68 595 (14.9%) | 55 283 (14.3%) | 21 271 (12.5%) |
| American, Western Europe total | 109 899 (23.9%) | 92 283 (23.9%) | 42 303 (24.9%) |
| Cyprus | 3[a] | 3 | — |
| Hungary | 140[a] | 137 | 67 |
| Poland | 532 | 506 | — |
| USSR | 64 204 | 62 339 | 34 068 |
| Czechoslovakia | 7 160 | 7 007 | — |
| USSR, Eastern Europe total | 72 036 (15.6%) | 69 989 (18.1%) | 34 135 (20.1%) |
| Total | 460 598 (100%) | 385 746 (100%) | 169 896 (100%) |

Source: IFR and National Robot Societies.
[a] Estimates by the ECE secretariat.
[b] Canada Ontario State Government (Sep. 85)

a serious labour shortage in the late 1960s – a decade which witnessed high economic growth in Japan. The gross national product showed an average 12% annual growth during this period. Skilled labour shortages became more acute and widespread, rising to 1.8 million in 1965. This eventually prompted efforts to develop industrial robots and to put them to practical use.

The first oil crisis of October 1983 forced Japan to shift its direction to low economic growth. The result of this was that the strain on the labour market was considerably eased. However, the country's higher oil and other natural resource costs drastically increased commodity prices and consequently labour costs. The fight against this cost inflation inevitably required improvements in productivity. Private businesses were encouraged to increase investment in automation.

Japan's population shows a 1.1% increase per annum. In contrast the growth rate of the labour force has been levelling off at 0.7%, mainly because of the growing number of students attending higher educational institutions. This has increased the need for higher productivity in order to achieve the annual economic growth target of 5%. The need is particularly felt in the manufacturing sector in which an increase in the labour force is not expected (Table 1.13).

The shortage of skilled labour has once again reached significant proportions owing to the tendency towards a highly educated society. Since November 1990, skilled labour shortages amount to about 3 007 000 workers (Table 1.14).

There has been a growing requirement to reduce the number of industrial accidents and occupational diseases due to hard and dangerous

**Table 1.13** Changes in the labor force by industry (×10 000)

| Industry | 1975 | 1980 | 1985 | 1989 | 1993 | 2000 |
|---|---|---|---|---|---|---|
| Primary | 661 | 577 | 509 | 463 | 439 | 345 |
| Secondary | 1841 | 1926 | 1992 | 2069 | 1970 | 1954 |
| Manufacturing | 1346 | 1367 | 1453 | 1484 | 1395 | 1335 |
| | (25.8%) | (24.7%) | (25.0%) | (24.2%) | (22.7%) | (21.2%) |
| Construction | 479 | 548 | 530 | 578 | 570 | 614 |
| Mining | 16 | 11 | 9 | 7 | 5 | 5 |
| Tertiary | 2721 | 3033 | 3306 | 3566 | 3725 | 4012 |
| | (52.0%) | (54.6%) | (56.9%) | (58.2%) | (60.7%) | (63.3%) |
| Total | 5223 | 5536 | 5807 | 6128 | 6134 | 6311 |
| | (100%) | (100%) | (100%) | (100%) | (100%) | (100%) |

Sources: 1975–1989, *Annual Report on the Labor Force Survey*, Statistics Bureau, Prime Minister's Office, Japan
1993, 2000, Economic Planning Agency survey

**Table 1.14** Shortage of skilled labor force in Japan (unit: 1000 person)

| | 1985 (June) | 1989 (Nov.) | 1990 (Nov.) |
|---|---|---|---|
| Total | 572 | 1908 | 3007 |
| Civil engineering | 89 | 367 | 688 |
| Manufacturing industry | 286 | 716 | 1071 |
| Wholesale, retail business, restaurant | 55 | 436 | 567 |
| Service industry | 96 | 221 | 450 |
| Others | 46 | 168 | 231 |

work and unfavourable working conditions. A great deal of pressure has been brought to bear on management to meet this need through the increased use of industrial robots.

A further factor stems from the progress of control technologies which are mainly based on microcomputers, progress in the field of robotics, and improved economic efficiency.

Japanese labour unions are not primarily trade unions, but are industrial unions, or unions organized on a company basis. This makes it easier to introduce industrial robots and to relocate workers.

### 1.5.4 THE PRESENT AND FUTURE STATUS OF INDUSTRIAL ROBOTS

The main Japanese industrial robot users in 1991 were as follows (Figure 1.14):

(a) electrical and electronic equipment manufacturers (whose share of the total Japanese distribution of industrial robots in 1991 was 41% in value);
(b) automobile manufacturers (25%);
(c) general machinery manufacturing (6%);
(d) plastics moulding and processing (6%);
(e) metal working (1%);
(f) precision machinery manufacturing (1%).

In addition, food, timber products, chemical, ceramics, steel making, and non-ferrous metal industries were also industrial robot users. This indicates that industrial robots are in demand by a wide range of Japanese industries. The demand for industrial robots in non-manufacturing industries, such as the nuclear power industry and the ocean development industry, only accounted for 2% of the total demand in 1991.

Table 1.15 shows the numbers of industrial robots supplied to major industrial processes in Japan for the years 1980, 1985, 1990 and 1991.

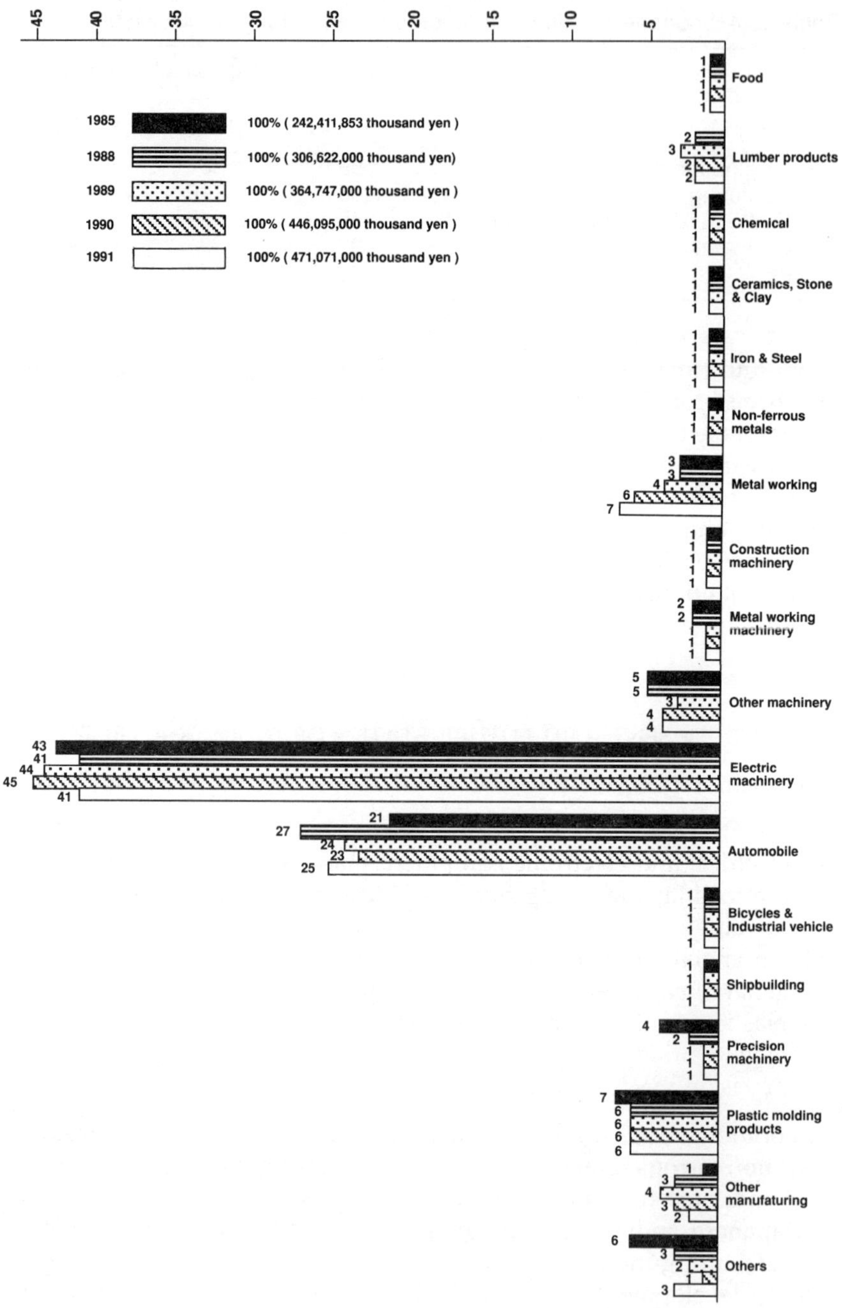


Figure 1.14 Distribution of industrial robots in Japan by industry (percentages), 1985–1991.

Table 1.15 Numbers of industrial robots in Japan, by industrial process (thousands)

| Process | 1980 | 1985 | 1990 | 1991 |
|---|---|---|---|---|
| Plastic molding | 7.7 | 9.5 | 12.9 | 12.3 |
| Machining | 3.1 | 3.9 | 6.3 | 6.6 |
| Assembly | 2.4 | 12.7 | 26.4 | 22.7 |
| Welding | 1.6 | 5.7 | 12.2 | 11.9 |
| Press | 0.8 | 1.1 | 1.8 | 1.7 |
| Casting | 0.7 | 0.9 | 1.7 | 1.8 |
| Inspection/Measurement | 0.1 | 0.7 | 0.9 | 1.1 |
| Painting | 0.1 | 0.4 | 0.9 | 1.0 |

Tables 1.16 and 1.17 show the distribution of industrial robots by industry and working processes for the past ten years.

It is anticipated that the automation of assembly processes for small batch production will become possible in different sectors of the machinery industry as the technological developments of intelligent robot systems progresses.

Exports stood at 21% of the total 1991 distributions and are expected to reach approximately 25% in the year 2000. In order for this to occur, Japanese manufacturers will need to collaborate with overseas engineering companies. The reason for this is that industrial robots require highly sophisticated systems engineering prior to installation, and good maintenance services after they are put into operation.

Apart from the manufacturing industry, wider use of robot technology and robotics is expected in the following sectors:

(a) nuclear – for safety inspections and maintenance of nuclear power plants, and radioactive waste disposal;
(b) medical social welfare – to help disabled people (2.4 million in 1987, of whom 1.46 are physically handicapped, and 0.3 million are visually handicapped) work more efficiently, and to help the bedridden (0.7 million in 1984) to earn a living;
(c) ocean exploration – for underwater construction, machining and geological surveys under water, and multi-purpose observations;
(d) agriculture and forestry – for fruit harvesting, crop dusting, lumbering, trimming, and timber collection;
(e) construction – for assembly of reinforced bars, painting bridges, and interior and exterior finishing of high-rise buildings;
(f) metal mining and coal mining;
(g) transportation and other service sectors.

**Table 1.16** Number of robots installed in Japan at the end of 1988

| 3.2 Application | | Types | | | | Total 2.1 |
|---|---|---|---|---|---|---|
| | | Sequenced robots 3.3 | Trajectory oper. robots 3.3 | Adaptive robots 3.3 | Teleoperated robots 3.3 | |
| 110. Handling operation for casting | | | | | | |
| 111. Mould | | (4 154) | (561) | (63) | (3) | |
| 119. Others | | (1 622) | (333) | (0) | (4) | |
| | Total | 5 776 | 894 | 63 | 7 | 6 740 |
| 130. Handling operation for plastic moulding | | 35 491 | 8 869 | 20 | 0 | 44 380 |
| 140. Handling operation for heat treatment | | 28 | 84 | 2 | 4 | 118 |
| 150. Handling operation for stamping / forging | | 2 794 | 1 221 | 0 | 1 | 4 016 |
| 160. Welding | | | | | | |
| 161. Arc welding | | (853) | (44 454) | (794) | (0) | |
| 162. Spot welding | | (109) | (24 721) | (21) | (0) | |
| 163. Gas welding | | — | — | — | — | |
| 164. Laser welding | | — | — | — | — | |
| 169. Others | | — | (347) | (3) | — | |
| | Total | 962 | 69 522 | 818 | 0 | 71 302 |
| 170. Dispensing | | | | | | |
| 171. Painting | | (66) | (5 413) | (1) | (0) | |
| 172. Sealing / Gluing | | (16) | (1 715) | (85) | (0) | |
| 179. Others | | — | — | — | — | |
| | Total | 82 | 7 128 | 86 | 0 | 7 296 |

| | | | | | |
|---|---|---|---|---|---|
| 180. Machining | | | | | |
| 181. Machine loading / unloading | (9 299) | (12 077) | (393) | (0) | |
| 182. Mechanical cutting / grinding / deburring / polishing | (82) | (2 576) | (80) | (3) | |
| 189. Others | (12) | (389) | (6) | (0) | |
| Total | 9 393 | 15 042 | 479 | 3 | 24 917 |
| 190. Other processes | | | | | |
| 191. Laser | | | | | |
| 192. Water jet | | | | | |
| 199. Others | | | | | |
| Total | | | | | — |
| 200. Assembling | | | | | |
| 201. Mechanical assembling / attachment | (6 523) | (73 529) | (8 916) | (0) | |
| 202. Inserting / mounting / cutting (e.g. PCB) | (560) | (19 658) | (3 793) | (0) | |
| 203. Bonding | (42) | (271) | (8 548) | (0) | |
| 204. Soldering | (34) | (1 001) | (0) | (0) | |
| 205. Handling for assembly operations | (384) | (5 953) | (123) | (5) | |
| 209. Others | | | | | |
| Total | 7 543 | 100 412 | 21 380 | 5 | 129 340 |
| 210. Handling operations for palletizing / packaging | 637 | 9 512 | 354 | 5 | 10 508 |
| 220. Measurement / Inspection / Testing | 417 | 3 198 | 1 661 | 54 | 5 330 |
| 230. Material Handling | | | | | |
| 240. Training / Education / Research | | | | | |
| 500. Others | 2 856 | 16 996 | 813 | 283 | 20 948 |
| 000. Unspecified | | | | | |
| | | | Grand total (equal to D 2.1 page 2) | | 324 895 |

Source: JIRA

**Table 1.17** Number of robots installed in Japan at the end of 1988

| 3.1 Industrial field | Types | | | | Total 2.1 |
|---|---|---|---|---|---|
| | Sequenced robots 3.3 | Trajectory oper. robots 3.3 | Adaptive robots 3.3 | Teleoperated robots 3.3 | |
| Manufacturing industries | | | | | |
| 31. Food / Beverages | 219 | 1 316 | 550 | 0 | 2 085 |
| 32. Textiles / Apparel / Leather products | 36 | 271 | 12 | 0 | 319 |
| 33. Wood / Wood products / Furniture | 66 | 2 258 | 0 | 0 | 2 324 |
| 34. Paper / Paper products / Printing and publishing | 162 | 634 | 3 | 0 | 799 |
| 35. Chemicals and Chemical, Petroleum, Coal, Rubber & Plastic products | 34 549 | 10 832 | 293 | 2 | 45 676 |
| 36. Ceramics / Stone products / Non metallic mineral products | 943 | 2 213 | 143 | 2 | 3 301 |
| 37. Basic metal | 342 | 954 | 188 | 15 | 1 499 |
| 381. Metal products (except machinery) | 2 777 | 12 187 | 322 | 0 | 15 286 |
| 382. Machinery (except electrical) | 3 269 | 28 053 | 490 | 3 | 31 815 |
| 383. Electrical equipment and apparatus | 10 096 | 85 148 | 20 646 | 0 | 115 890 |
| 384. Transport equipment | | | | | |
| 3843. Motor vehicles | (9 040) | (71 918) | (940) | (0) | |
| 3845. Aircraft | (—) | (7) | (2) | (0) | |
| Others | (392) | (4 064) | (107) | (20) | |
| Total | 9 432 | 75 989 | 1 049 | 20 | 86 490 |

| | | | | | |
|---|---|---|---|---|---|
| 385. Precision / Optical instruments | 1 461 | 5 092 | 1 281 | 0 | 7 834 |
| 39. Other manufacturing industries | 2 406 | 6 215 | 557 | 0 | 9 178 |
| | | Manufacturing industries total | | | 322 496 |
| Non manufacturing industries | | | | | |
| 1. Agriculture / Forestry / Fishing | | | | | |
| 2. Mining | | | | | |
| 4. Electricity / Gas / Water supply | | | | | |
| 5. Construction / Building | | | | | |
| 931. Education | | | | | |
| 932. Research / Development | | | | | |
| Other non-manufacturing industries | 221 | 1 716 | 142 | 320 | 2 399 |
| | | Non manufacturing industries total | | | 2 399 |
| Unspecified | | | | | |
| | | | Grand total (equal to D 2.1 page 2) | | 324 895 |

Source: JIRA (Japan Industrial Robot Assoc.)

### 1.5.5 INTERNATIONAL COOPERATION IN THE FIELD OF INDUSTRIAL ROBOTS

The improvement of productivity results in economic development. Common objectives worldwide are increased employment, the prevention of industrial accidents due to hazardous and unfavourable working conditions, the improvement of labour welfare, and improved working environments. These are the reasons for international cooperation in the field of industrial robotics. A remarkable progress is anticipated in the future.

The decision on 'International Research and Cooperation in the Field of Advanced Robotics' made at the summit meeting of the seven industrialized democracies in Williamsburg, USA, May 1983, shows that industrial robots are now rated as a major theme for international industrial and research cooperation.

#### (a) International exchange of information on research and technology

On 30 September 1986 at the 16th International Symposium on Industrial Robots (ISIR) in Brussels, Belgium, the national coordinators' meeting agreed on the creation of an International Federation of Robotics (IFR). The IFR was established on 1 January 1987 with 15 members (see Appendix 1.A). It now plays a central role in international robotics cooperation and also organizes the ISIR and the International Conference on Advanced Robotics (ICAR).

Remarkable progress has been made in international cooperation in the field of robotics. This is mainly through the exchange of research and technological information, centred around ISIR. Such cooperation will be further encouraged in the future.

The first Symposium on Industrial Robotics was held in the USA in 1970. A decision was made to turn this into an international event. The ISIR has since been held in the USA, Europe and Japan, with between 300 and 1000 delegates from about 20 countries attending each time, presenting more than 50 papers. This has greatly contributed to the development of industrial robots.

At previous meetings of the ISIR, a large number of papers have been presented by various countries on research and development, applications, and the socio-economic impact of the use of industrial robots. This has opened the path for the ISIR to build a foundation for an international exchange of research and technological information (see Appendix 1.B).

The summit of seven industrialized countries at Versailles, France, in June 1982 produced a proposal regarding international cooperation in research on advanced technologies. The subject of advanced robotics was adopted at the Williamsburg summit meeting of May 1983 as a theme for international research cooperation. Japan and France were named as the leading countries on this project.

The Japanese Industrial Robot Association (JIRA) held the first International Conference on Advanced Robotics (ICAR) in Tokyo, September 1983. This conference was held with a view to an international exchange of research information and to present the results of research and national projects on advanced robotics (see Appendix 1.B).

### (b) International industrial cooperation

Japanese international industrial cooperation on industrial robots is being promoted by technological seminars for industrial cooperation overseas, international contracts for marketing and technology, and the establishment of joint ventures.

Technology seminars have so far been held at the request of foreign countries under the sponsorship of the Japanese External Trade Organization (JETRO), with the cooperation of JIRA (see Table 1.18). The main topics of discussion have included the technology of industrial robots, their current and future situations, and the social and economic impact of robot utilization. These seminars have brought about many favourable results, including the establishment of international joint ventures and progress in international industrial cooperation.

JETRO has dispatched investment missions to southeast Asia (Singapore and Malaysia) and Latin America (Mexico and Brazil). With JETRO's help, Japan (through the Japanese Leasing Company) and Singapore (EOB, SAL) have launched a joint venture, the Robot Leasing and Consultancy Pty Ltd in Singapore.

In contrast to cars and domestic electric appliances, industrial robots are not easy to export. It is necessary to provide a systems engineering service prior to introducing industrial robots. In addition, guaranteed maintenance needs to be provided after introduction. The requirements for production systems vary considerably from one user to another. This is due to differences in product types (shape, size and eight), the number of production lots, plant layout, cycle times, and positional accuracy requirements. Systems engineering involves making a basic design of the production system by adjusting these requirements, performing a feasibility study using robots, and selecting the most suitable robots for the task. Japanese manufacturers have to satisfy these user demands and requirements by collaborating with foreign systems engineering companies or manufacturers who can supply the necessary systems engineering services. This results in the provision of imported robots with high added value and maintenance services to end users. Cooperative relationships between Japanese manufacturers and their overseas competitors have therefore been set up, from the beginning, without generating trade friction.

**Table 1.18** Technology seminars on industrial robots, sponsored by JETRO with the cooperation of JIRA

| Name of seminar | Venue | Date |
|---|---|---|
| Symposium associated with an exhibition of Japanese industrial robots in Moscow | Moscow | 11 & 14 April 1975 |
| Technology seminar on industrial robots in Eastern Europe | | |
| Czechoslovakia | Bruno | 17 November 1978 |
| | Prague | 21 November 1978 |
| | Cosiche | 23 November 1978 |
| Bulgaria | Stara Zagora | 28 November 1978 |
| Technology seminar associated with engineering show in Sydney | Sydney | September 1978 |
| Technology seminar in Australia | Canberra | September 1980 |
| | Melbourne | October 1980 |
| | Sydney | October 1980 |
| Seminar on industrial robots in Paris | Paris | March 1981 |
| Technology seminar for Europe – Japan industrial cooperation | | |
| Italy | Milan | September 1981 |
| France | Paris | September 1981 |
| UK | London | September 1981 |
| Technology seminar on industrial robots in Canada | Toronto | 22 February 1982 |
| | Vancouver | 24 February 1982 |
| | British Columbia | 25 February 1982 |
| Seminar for industrial cooperation in North America | Detroit | 26 October 1982 |
| | Chicago | 28 October 1982 |
| | Minneapolis | 2 November 1982 |
| | Purdue University | 29 October 1982 |
| Technology seminar for Europe – Japan industrial cooperation | | |
| Sweden | Stockholm | 10 October 1984 |
| Austria | Wien | 12 October 1984 |
| Spain | Madrid | 16 October 1984 |
| UK | London | 18 October 1984 |
| Seminar for PRC – Japan industrial robot cooperation | Wuhan | 31 October 1986 |
| | Shanghai | 4 November 1986 |
| | Beijing | 6 November 1986 |
| Japan–UK industrial technology cooperation seminar | London | 18 September 1991 |

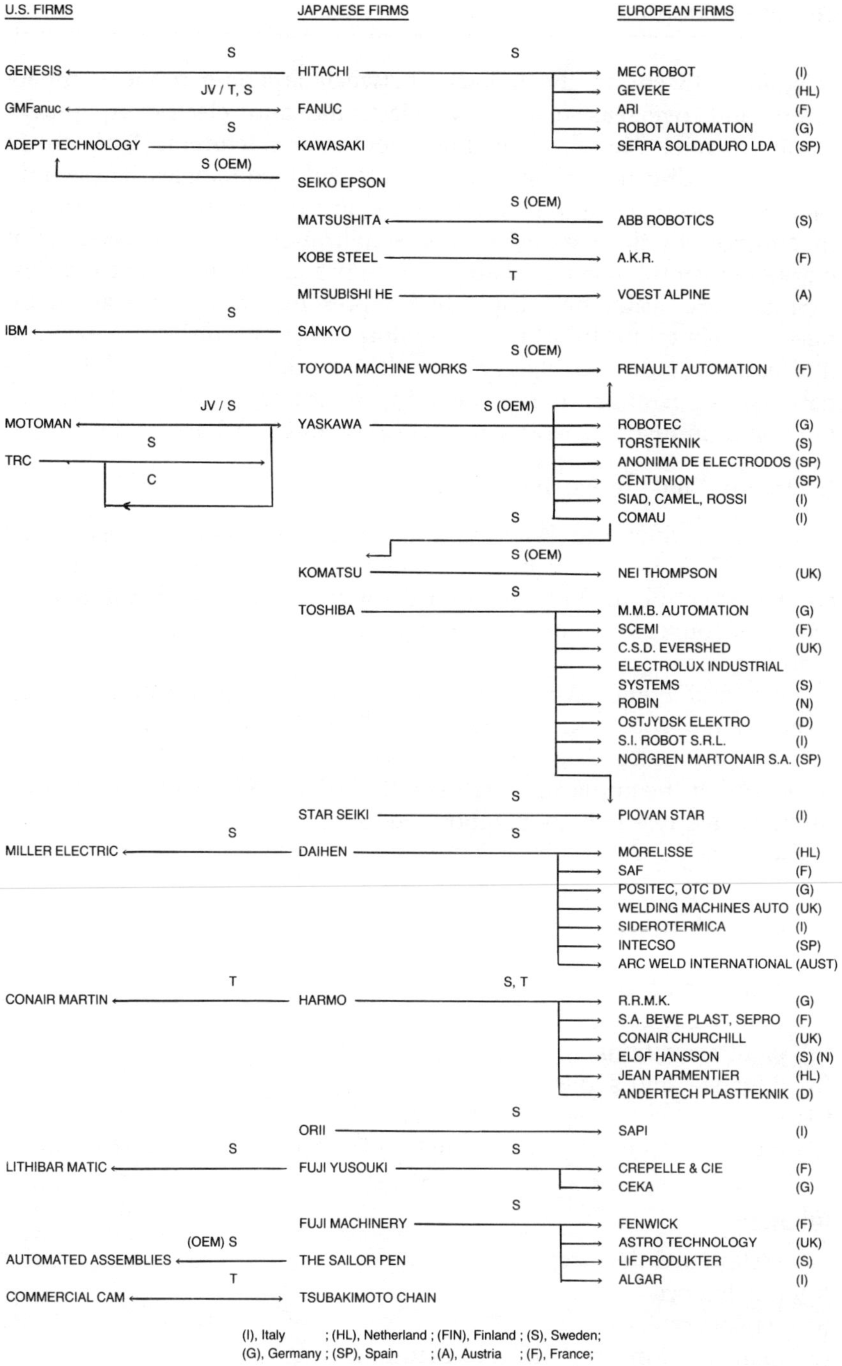


**Figure 1.15** International industrial cooperation on robots (April 1992). T, technical links; S, sales links; JV, joint ventures. Arrows indicate the flow of products and technologies.

Figure 1.15 shows the linkages between Japanese robot manufacturers and overseas machinery, electronic and electric equipment manufacturers. These are mainly robot manufacturers. The arrows indicate the direction of flow of products and technologies. These linkages have been set up mainly at the request of foreign manufacturers. In addition to the expansion and stabilization of the market, joint marketing ventures will probably shift to production in other countries.

Since 1984, JIRA has dispatched representatives in charge of industrial robots to JETRO's Düsseldorf office in order to expedite international industrial cooperation. JIRA has also contributed to international cooperation in this field by mounting an international exhibition of industrial robots whenever ISIR and ICAR are held in Japan.

**(c) International research cooperation**

It was agreed at the Williamsburg summit of seven industrialized nations, USA 1983, that international research cooperation should be conducted in the field of advanced robotics. It was considered that this would encourage economic growth and an improvement in the labour market.

Japan and France were appointed as the leading countries in this international cooperation. The Joint Coordinating Forum (JCF) for international research cooperation was held in January 1984. The JCF decided that the application fields of the Advanced Robotics Project for international research cooperation are as follows:

(a) nuclear power;
(b) underwater operations;
(c) firefighting and rescue operations;
(d) agriculture;
(e) mining and tunnelling;
(f) civil engineering;
(g) plant operations;
(h) domestic and service;
(i) space;
(j) medical and health care (added at the April 1986 JCF in Hawaii).

The elemental technologies for the Advanced Robotics Project are as follows:

(a) system architecture;
(b) intelligence;
(c) sensors;
(d) man–machine systems and manipulation;
(e) manipulation;
(f) locomotion;
(g) actuators.

The need for economic development and employment stability is a worldwide problem. Consequently it can be predicted that international cooperation into advanced robotics will increase greatly. Meanwhile, in Japan, MITI launched an eight-year national research project on 'advanced robot technology' in 1983. This was aimed at developing robots capable of operating in extreme and hazardous environments, e.g. nuclear power plants, offshore oil exploration and disaster prevention.

This national project is playing its role as the Japanese core international advanced robotics project. In addition, 20 Japanese private organizations established the Advanced Robot Technology Research Association (ARTRA) in February 1984. In June 1985 Japan established the International Robotics and Factory Automation Centre (IROFA) to carry out international cooperative research in robotics. It also implements national joint research amongst the private sector, national institutes and universities.

### 1.5.6 ECONOMIC AND SOCIAL IMPACT OF INDUSTRIAL ROBOT INSTALLATION

Productivity must be improved in order to achieve stable economic growth against cost-push inflation. This is especially true in Japan, which lacks natural resources and therefore has to support her national economy by means of trade. There is a need to prevent industrial accidents and occupational diseases and to improve the quality of human labour. In particular there is a need to reverse the trend of dehumanization, which is turning mankind into a slave of the machine as exemplified by the conveyor system. The provision of an agreeable working life represents an important key for the promotion of industrial and social welfare.

Industrial robots provide a powerful means of achieving these objectives. Their application is wide-ranging in terms of economic and social impact, the reason for this being their particular characteristics, such as flexible movement functions, which enable the automation of small batch production. They change production from man–machine based to man–robot–machine based. The introduction of industrial robots removes workers from dangerous and monotonous tasks, and allows the upgrade of the quality of human labour to activities such as teaching, operating and maintenance.

The reasons stated are sufficient to justify the need for industrial robots. However, they not only improve productivity and welfare, but also upgrade product quality and effects savings in resources and energy. This makes capital investments more effective, reduces rejection rates, improves production control, stabilizes employment, solves

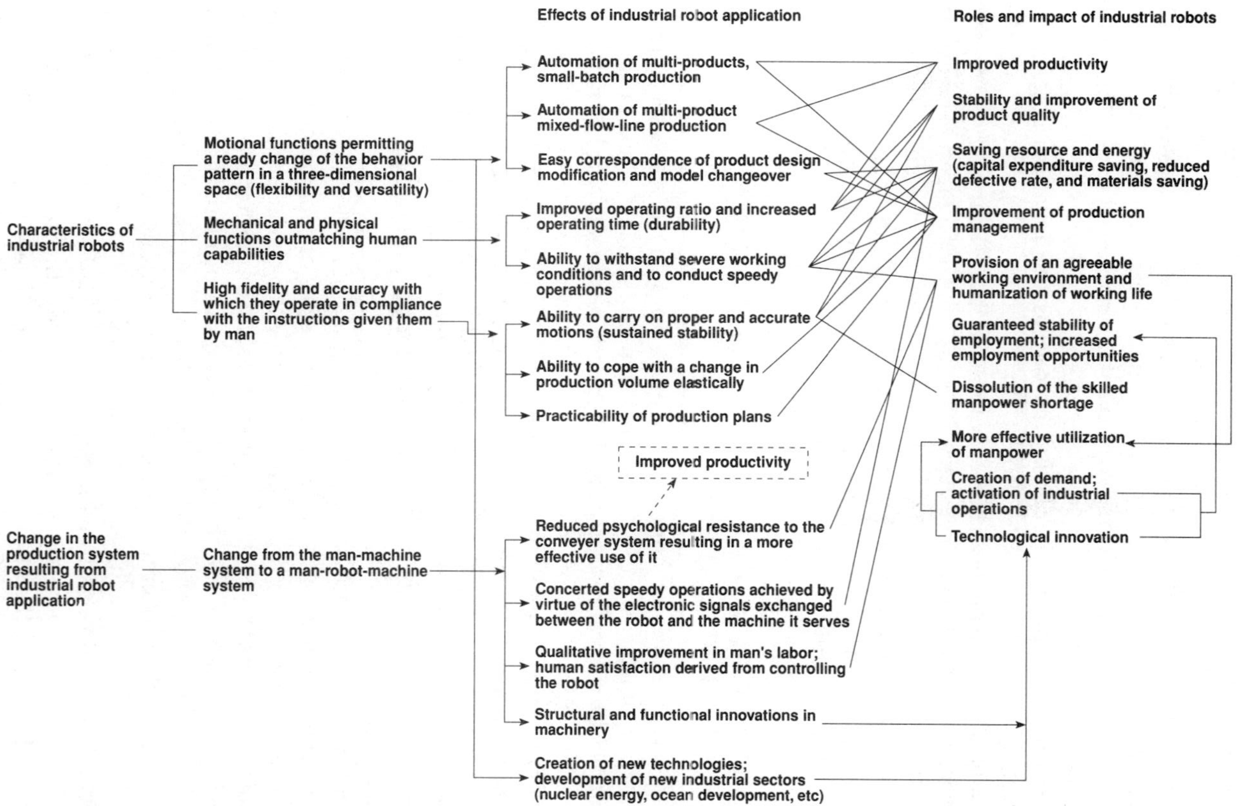


Figure 1.16 Socioeconomic and technological impact of industrial robots.

manpower shortage problems, creates new technologies, and enables the development of new industrial sectors.

Figure 1.16 illustrates the advantages resulting from industrial robots' operational characteristics and the resultant changes in production systems. It is these advantages which account for the large impact of industrial robots, and why they are expected to play an increasingly important role in the future.

The reasons and significance of the impact of industrial robots can be more specifically explained as follows.

### (a) Improved productivity

In contrast to more specialized automatic machines, industrial robots can readily be reprogrammed to perform different tasks, or variations on a particular task. The ability to meet changing requirements in operating mode is referred to as time elasticity. The ability to accommodate variations on a particular task is referred to as space elasticity. These characteristics permit the automation of multi-product, small batch or multi-product, mixed flow production. For this reason industrial robots are considered as advanced automation equipment.

Industrial robots are capable of continuous operation without interruption. This helps to greatly improve production. The introduction of industrial robots changes a man–machine system into a man–robot–machine system, where robots service conveyors directly in place of human operators.

The conveyor system is an effective means of promoting productivity. It does however have a major disadvantage in terms of its dehumanizing effect on the workforce – this was highlighted in Charlie Chaplin's critical film 'Modern Times', released in 1936. This effect was taken up by the European Commission in 1973 when it proposed 'the total elimination of belt production' as an important labour policy for member nations. The introduction of industrial robots allows the improvement of productivity while eliminating the dehumanizing effect.

Improved productivity is required to strengthen international competitiveness and corporate profitability. The diversity of social needs are in turn diversifying the product mix. In Japan's machinery industry, 80% of total output derives from multi-product small or medium batch production, the automation of which leads to a drastic improvement in productivity. This improved productivity is also required to match the tendency towards wage inflation.

Improved productivity is also required to meet needs arising from the supply and demand situation of the labour force. Efforts are being made towards shorter working hours. However, Japan is tending towards a highly educated society. The result of this will be increasing difficulty in recruiting blue collar workers. The anticipated growth rate of the labour force in the coming years is as low as 0.7%. It is expected that the percentage of the overall working population represented in the tertiary industry, which was 56.9% in 1985, will gradually increase to 63.3% in 2000. This is following the pattern of the USA which has already reached about 70%. Since it cannot be expected that a significant growth of the secondary industry workforce will occur, improvements in productivity will play an important role in the development of this sector.

### (b) Provision of agreeable working environment and the humanization of working life

Industrial robots are currently taking over dangerous, severe, trying and monotonously repetitive tasks in a gradual manner. These tasks are only performed by human beings because of difficulties in achieving automation. This trend will go a long way towards the prevention of the industrial accidents and occupational diseases which result from such unfavourable jobs.

It is expected that at the same time, the human workers will be turning towards the high quality jobs needed to sustain automation technology. These jobs include robot operators, maintenance personnel, and instructors for training people for these roles. Workers may also be employed as systems engineers designing aspects of new robot-assisted production systems, and as technical staff developing new industrial robots. These moves are important considerations for the promotion of industrial welfare and humanization. This improvement in the quality of human labour can provide an effective means of coping with the trend towards a growing number of white-collar workers on the shop floor, a trend which is being observed more and more in Japan as society becomes increasingly educated.

The increased utilization of robots can also play an important role in promoting the employment of older people. This is particularly significant in societies such as Japan which has an increasing average age of population.

Efforts are being made to develop robots for use in caretaking and firefighting applications. They can reduce the workload of nursing staff caring for the bedridden, and can aid the handicapped to be gainfully employed. They could also take over the role of guide dogs serving the visually handicapped.

### (c) Stability of employment and increased employment opportunities

Dangerous, severe, trying and monotonous work is disliked by workers and leads to high staff turnovers. Industrial robots can be employed to perform unfavourable jobs with a view towards improving the quality of work and promoting a greater workforce stability. This leads to greater employment opportunities for white-collar and older workers.

### (d) Promoting the development of new industries

Ocean development can be promoted by using industrial robots for underwater operations which are difficult for human workers to perform. It is also possible to facilitate the development of nuclear power by using robots for reactor operations and maintenance where the risk of radiation injury is greatest. There are great expectations for the use of robots in these industrial sectors. Progress in these sectors will also promote economic growth, employment opportunities, and improved manpower utilization.

### (e) Increased return on capital investment

Night working is generally unpopular, hence single-shift day working is currently common practice. There are exceptions to this in special cases such as the steel-making and automobile industries. In the man–robot–machine system, as opposed to the conventional man–machine system, 'man' represents a minority activity which is more intellectual than in the conventional system. The resulting freedom from physical activity and endurance enables a two- or three-shift working system to be employed. The consequence of this is that it is possible to reduce costs of a given output by a factor of two or three. The net result of this cost saving is an increased return on capital investment, and resource conservation.

Even when automated mass production is achieved by means of special purpose machines, changing economic and social needs necessitate design modifications and model changes. It is very expensive and time consuming to accommodate such changes, modifying existing machines and equipment. In extreme cases machines may need to be completely replaced. In contrast, when using industrial robots, many design modifications and model changes can be achieved simply by reprogramming.

### (f) Other impacts

Other impacts brought about by the utilization of industrial robots are:

(a) improved stability of product quality;
(b) resource conservation resulting from reduced rejection rates and savings in materials;

(c) improved product control;
(d) a solution to the problem of skilled labour shortages.

### 1.5.7 PROSPECTIVE ROLES FOR INDUSTRIAL ROBOTS

Finally, the roles which industrial robots are expected to play in the future can be summarized as follows:

(a) economic growth sustained by improved productivity and the resulting increase in jobs;
(b) contributing to the welfare of workers, and eliminating industrial accidents;
(c) solving the problem of skilled labour shortages;
(d) meeting the need for highly intellectual duties within factories and other establishments;
(e) contributing to the growth of new industries, including nuclear and ocean exploration, and the creation of new jobs in these areas;
(f) contributing to an increase in employment opportunities for the physically handicapped and to social welfare by assisting the bedridden.

## APPENDIX 1.A ROBOT ASSOCIATIONS AND SOCIETIES THROUGHOUT THE WORLD

IFR member associations are indicated by an asterisk (*).

Japan Industrial Robot Association (JIRA)*
Kikaishinko Building, 3-5-8, Shiba Kohen, Minato-ku, Tokyo 105, Japan
Established: October 1972

Robotic Industries Association (RIA)*
PO Box 3724, Ann Arbor, MI 48106, USA
Established: June 1974

Italian Society for Industrial Robots (SIRI)
(Association Italiana di Robotica)
c/o ETAS Periodici, Via Mecenate 91 1-20138 Milano, Italy
Established: 1975

British Robot Association (BRA)*
Aston Science Park, Love Lane, Aston Triangle, Birmingham B7 4BJ, UK
Established: 1977

French Industrial Robot Association (AFRI)
(Association Français de Robotique Industrielle)
Tour 66, 4, Place Jussieu, F-75252 Paris Cedex 05, France
Established: 1978

Robotics International of SME
One SME Drive, PO Box 930, Dearborn, MI 48121, USA
Established: 1980

Swedish Industrial Robot Association (SWIRA)*
Sveriges Makanforbund, Box 5506, S-11485 Stockholm, Sweden
Established: 1980

Belgian Institute of Automatic Control (BIRA)-Robotica*
(Belgisch Instituut voor Regeltechniek en Automatizering VZW)
Desguinlei 214, B2018 Antwerpen, Belgium
Established: 1980

Australian Robot Association Inc.*
GPO Box 1527, Sydney 2001, NSW, Australia
Established: 1981

Fachgemainschaft montagetechnik Hanbdhabungstechnik und Industrie Robot (MHI), VDMA*
Lyoner Strasse 18, Postfach 710864, D-6000 Frankfurt (Main) 71, Germany
Established: October 1981

Danish Industrial Robot Association (DIRA)*
(Dansk Robot Forening)
Teknologiparken, DK-8000 ARHUS C, Denmark
Established: February 1982

Singapore Industrial Automation Association (SIAA)*
151 Chin Swee Rd #03-13 Manhattan House, Singapore 0511, Republic of Singapore
Established: June 1982

Singapore Robot Association
Maxwell Road, PO Box 392, Singapore 9007, Republic of Singapore

Flemish Organization of Robotics and Automation (FLORA)
Brouwersvliet 15 bus 7, B-2000 Antwerpen, Belgium
Established: September 1982

Robotics Society of Japan
c/o Bunkyo-Shogaku Building, 1-15-4, Hongo, Bunkyo-ku, Tokyo 113, Japan
Established: January 1983

The Robotics Society of Finland
Asemapaallikonkatu 12C, BF-00520 Helsinki, Finland
Established: March 1983

Association de Robotique Industrielle, Athena, a.s.b.l.
Centre de Services de Lauzelle, Avenue de Lauzelle, 61, 1348 Ottingnies-Louvain-la-Neuve, Belgium
Established: March 1984

The National Service Robot Association (NSRA)
c/o Robotic Industries Association, USA
Established: August 1984

Spanish Association of Robotics (AER)*
(Association Espanola de Robotica)
Rambla de Catalunya, 70, 3 2a, E-08007, Barcelona, Spain
Established: May 1985

International Robotics and Factory Automation Centre (IROFA)
Daiichi Nakano Building, 2-6-10, Iwamoto-cho, Chiyoda-ku, Tokyo 101, Japan
Established: June 1985

Dutch Robotic Association (CIR)
(Contactgroep Industriele Robots)
c/o Hoogovens IJmuiden BV Research Lab (3J22), 1970 CA IJmuiden, The Netherlands

Hungarian Robotics Association*
PO Box 433, H-1371 Budapest, Hungary

International Federation of Robotics (IFR)*
Mekanforbundet PO Box 5506, S-114 85 Stockholm, Sweden
Established: January 1987

Ostereiches Forschungszentrum Seibersdorf GmbH*
Hauptabteilung Fertigungstechnik und Automation, A-2444 Seibersdorf, Austria

China Society of Industrial Automation and Automated Industry*
5th Floor, 106 Hoping E Road, Sec 2 Taipei, 106 36 Taiwan, Republic of China

IPA*
Nobelstrasse 12, D-7000 Stuttgart 80, Germany

Meininger Automation bv*
Postbus 743, NL-2280 AS Rijswijk, The Netherlands

Schweizersche Gesellschaft für Automatik*
Arbeitsgruppe Robotik, Postgasse 17, CH-3011 Bern, Switzerland

Federation of Norwegian Engineering Industries (MVL)*
Box 7072-H N-0306, Oslo 3, Norway

Polish Federation of Engineering Association (NOT)*
Czackiego Str 3/5. P-00950, Warsaw, Poland

* 'Robot' Interindustry Research and Production Association (MNTK 'Robot') Izmailovskaya Sq.7, Moscow, Russia, CIS

## APPENDIX 1.B INTERNATIONAL COOPERATION ON SCIENTIFIC AND TECHNICAL INFORMATION EXCHANGE ON ROBOTICS

International Symposium on Industrial Robots (ISIR)

| | |
|---|---|
| 1st April 1970 | Chicago, USA |
| 2nd May 1972 | Chicago, USA |
| 3rd May 1973 | Zurich, Switzerland |
| 4th November 1974 | Tokyo, Japan |
| 5th September 1975 | Chicago, USA |
| 6th March 1976 | Nottingham, UK |
| 7th October 1977 | Tokyo, Japan |
| 8th May 1978 | Boeblingen, Stuttgart, FRG |
| 9th March 1979 | Washington, USA |
| 10th March 1980 | Milan, Italy |
| 11th October 1981 | Tokyo, Japan |
| 12th June 1982 | Paris, France |
| 13th April 1983 | Chicago, USA |
| 14th October 1984 | Gothenburg, Sweden |
| 15th September 1985 | Tokyo, Japan |
| 16th September 1986 | Brussels, Belgium |

| | |
|---|---|
| 17th April 1987 | Chicago, USA |
| 18th April 1988 | Lausanne, Switzerland |
| 19th November 1988 | Sydney, Australia |
| 20th October 1989 | Tokyo, Japan |
| 21st October 1990 | Copenhagen, Denmark |
| 22nd June 1991 | Detroit, USA |
| 23rd September 1992 | Barcelona, Spain |
| 24th November 1993 | Tokyo, Japan |
| 25th April 1994 | Hannover, Germany |
| 26th October 1995 | Singapore |
| 27th October 1996 | Rome, Italy |

International Conference on Advanced Robotics (ICAR)

| | |
|---|---|
| 1st September 1983 | Tokyo, Japan |
| 2nd September 1985 | Tokyo, Japan |
| 3rd October 1987 | Versailles, France |
| 4th June 1989 | Columbus, Ohio, USA |
| 5th June 1991 | Pisa, Italy |
| 6th June 1993 | Tokyo, Japan |

# Manufacturing systems technologies and CIM 2

*Satoru Takashima*

The term computer-integrated manufacture (CIM) is used to describe advanced systems which embrace automation, systems engineering, computer science, mechatronics and standardization. A successful implementation of CIM is based upon the integration of a number of technologies, including the following:

(a) factory automation (FA);
(b) computer-aided design and computer-aided manufacturing (CAD/CAM);
(c) computer-aided engineering (CAE);
(d) communications (including telecommunications);
(e) human communication;
(f) the management of information and products;
(g) the control of information, products, materials and machines in a manufacturing environment.

Manufacturing systems can be designed using either top-down or bottom-up methods. Japanese manufacturing systems using FA and flexible manufacturing system, FMS, have adopted the latter approach.

The introduction of practical FA systems has been remarkably rapid in Japan. However, in contrast, there are currently few complete CIM systems. This chapter concentrates on the engineering applications of CIM and its enabling technologies as used in Japan. The research and development of CIM implementations for selected industries are also discussed. These implementations include the use of computers, controllers, computer numerial control (CNC) machine tools, robots and robotic systems (including automated guided vehicle systems).

*Manufacturing Automation Systems and CIM Factories*. Edited by K. Asai, S. Takashima and P.R. Edwards. Published in 1994 by Chapman & Hall, London. ISBN 0 412 48230 4

## 2.1 PROGRESS TOWARDS CIM

The 1980s saw an increase in the popularity of factory automation systems using robots, NC machines, automatic transfer machines and automatic transfer systems. Although the aim is a fully integrated manufacturing system, little progress has been made beyond semi-integrated systems. A significant number of CIM systems are still largely controlled by hardware-based control systems such as programmable logic controllers (PLC). The progress of manufacturing systems towards CIM can be measured by examining the phases of automation.

The first three phases use a single automatic control system. Manufacturing systems which utilize multiple software-driven control systems are described under phases four and five.

### 2.1.1 THE FIRST PHASE OF AUTOMATION

The first phase of the automation of mass-production manufacturing had the objective of improving productivity. During this phase transfer-line manufacturing systems were automated using conveyors.

### 2.1.2 THE SECOND PHASE OF AUTOMATION

The second phase of automating manufacturing systems was the integration of NC machines with transfer lines to provide flexibility. A flexible transfer line was required to accommodate the requirements of product variety and diversity. During this phase, NC machines developed into highly efficient, high functionality tools. Industrial robots began to provide a useful role in factories. Developments of computer-assisted systems which began in the late 1960s started to produce practical results in the mid 1970s.

### 2.1.3 THE THIRD PHASE OF AUTOMATION

The late 1970s saw the research and development of electronically assisted mechanical machines in Japan. These machines, known as mechatronics, have a high level of functionality and performance. Typical applications of mechatronics are industrial robots and NC machine tools. The development of mechatronic machines led to the construction of new automated manufacturing systems. These included FMS and flexible manufacturing cells (FMC), which were the integration of the following:

(a) automatic warehousing systems;
(b) automatic machine tools;
(c) automatic assembly machines and systems;

(d) automatic loading and unloading machines;
(e) automatic inspection and checking machines;
(f) automatic transfer machines;
(g) automatic materials handling systems.

At the same time, practical automatic NC programming systems (automatically programmed tools, APT) and languages were being developed which were incorporated into CAD/CAM systems. The developments in mechatronics and programming systems comprised the third phase of automating manufacturing systems.

These factory automation manufacturing systems require the information to control physical distribution systems, manufacturing activities and the progress of product manufacture. These systems form the basis for CIM.

### 2.1.4 THE FOURTH PHASE OF AUTOMATION

In the early 1970s, information-controlled manufacturing systems used direct numerical control (DNC). Later, computer numeric control (CNC) systems began to be used in manufacturing. Research and development of CAD, CAM, computer-aided engineering (CAE), computer-aided testing (CAT), and other off-site activities related to manufacturing systems also began in the early 1970s. CAD systems helped designers to create electronic elements and the mechanical shapes of parts and products. CAM systems helped designers and engineers to create manufacturing databases, process planning, and manufacturing programmes. CAE systems interconnected manufacturing shops, processes and cells. CAT systems helped engineers and workers to measure and inspect products.

From the early 1980s, computer-controlled systems came into general use in manufacturing systems as more powerful, more compact and lower priced computers became available. Sensors, which are a very important element of manufacturing control systems, improved in quality and performance characteristics. This led to adaptive control systems, the fourth phase of the automation of manufacturing systems. This enabled the automation of batch systems, which manufacture small quantities of a range of products, to satisfy the need for product variety in the current Japanese market. In the mid 1980s, CAD/CAM, CAE and CAT were in practical use by engineers and workers in manufacturing.

### 2.1.5 THE FIFTH PHASE OF AUTOMATION

The emphasis of automation in manufacturing systems changed from hardware to software. The construction of automated manufacturing systems is also dependent upon computer and communications technologies. In Japan, manufacturing systems are still based on human communication, for which Japanese industries have been very successful. In future Japanese manufacturing systems, CIM will be the norm with factory-based local area networks (LAN) providing communication. This is the fifth phase of the automation of manufacturing systems.

### 2.1.6 FUTURE DEVELOPMENTS

Current research ideas include intelligent machines and systems (see Figure 2.1). The next generation will aim at intelligent manufacturing management (IMM) systems which implement the results of research and development into expert systems, fuzzy systems etc. The phases of development are summarized in Table 2.1.

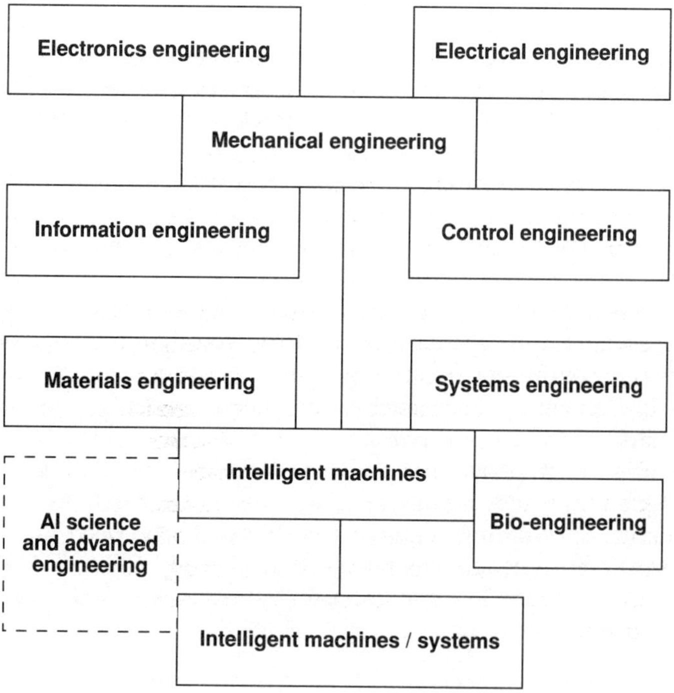


Figure 2.1 Current areas of research into intelligent machines and systems engineering (classification, growth, architecture, drive mechanisms, materials, communication, language, intelligence, standardization).

Table 2.1 History of automation in manufacturing systems

| Phase | Development | Period |
|---|---|---|
| 1 | Transfer line for mass production | Late 1960s |
| 2 | NC machines on a flexible transfer line | Middle 1970s |
| 3 | Mechatronics used in multi-production | Late 1970s |
| 4 | Computers for adaptive-controlled systems | Middle 1980s |
| 5 | Integrated management system networks | Middle 1990s |
| 6 | Artificial intelligence system networks | Early 2000s |

## 2.2 ROBOTICS AND ROBOT SYSTEMS ENGINEERING

Industrial robots are the key to standardization in CIM. Their essential elements include mechanisms, computers, sensors, interfaces and communications. There are three types of robot: manufacturing robots, service robots, which are practical robots working in a range of applications, and robots in their experimental or developmental stages.

## 2.3 JAPANESE INDUSTRIAL ROBOTS

Japanese robotics studies have focused on communications, intelligence, actuation power, architecture, standardization and real-time processing.

The history of Japanese robots can be considered to have begun with the Karakuri-Ningyoh puppets which were created in the Edo era. Japanese research and development into industrial robots began in 1967 following the introduction of the Unimate and Versatran robots in 1962. Robots have since been almost exclusively used in manufacturing systems. Practical robots have been utilized in factories. National robotics research and development projects have aimed at practical applications of automation and the improvement of the working environment. More recently, the use of robots has extended to architecture, oceanography and nuclear power. Current research and development is examining the potential use of robots in space.

New technologies, materials, architectures, control systems and communications systems are required to construct advanced robots. The new technologies include high-level engineering, sensors, computer science, artificial intelligence, multi-discipline technologies and integrated systems (see Figure 2.2).

### 2.3.1 ROBOTICS IN MANUFACTURING

In Japan, industrial robots have replaced human machine operators. This has resulted in increased productivity in automobile, electrical/

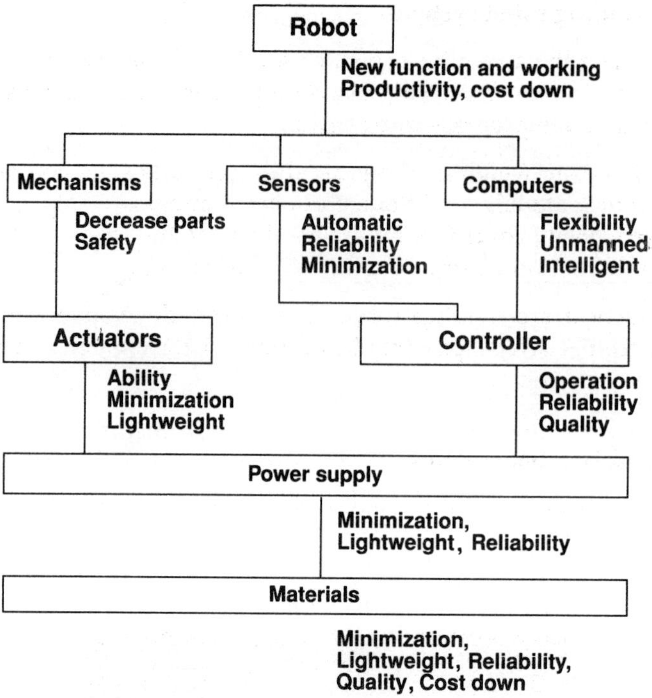


Figure 2.2 Configuration and functions of a robot.

electronics and machinery factories. Manufacturing robots have been used as far back as 1980 for loading and unloading, palletization, line transfer, spot welding and simple parts assembly. Today, the flexibility and dexterity of manufacturing robots has been extended to such diverse applications as arc welding, paint spraying, pick-and-place operations, practical assembly operations, inspection and the deburring of metal parts.

Industries require more sophisticated manufacturing systems to meet the advanced engineering and the use of automation and mechatronics in factories.

CIM and integrated intelligent manufacturing management (IIMM) systems also make use of software and have special needs in communications, programming and processing.

### 2.3.2 ROBOT APPLICATIONS IN CIM

The use of industrial robots in popular in Japanese factories. A representative selection of existing applications are described below.

## (a) Arc-welding robot systems

Arc-welding robots require a greater level of flexibility and dexterity than spot-welding robots. The following problems need to be solved in order for tangible progress to be made:

(a) decisions on welding methods and techniques to accommodate different material, conditions and environments;
(b) stable velocity control of an arc-welding torch;
(c) accurate motion control of an arc-welding torch.

Many practical arc-welding robots have been developed (see Figure 2.3) and are in use in Japan thanks to the development of an automatic arc-welding machine.

## (b) Assembly robot systems

Assembly robots have been used in the electrical/electronics and automobile industries.

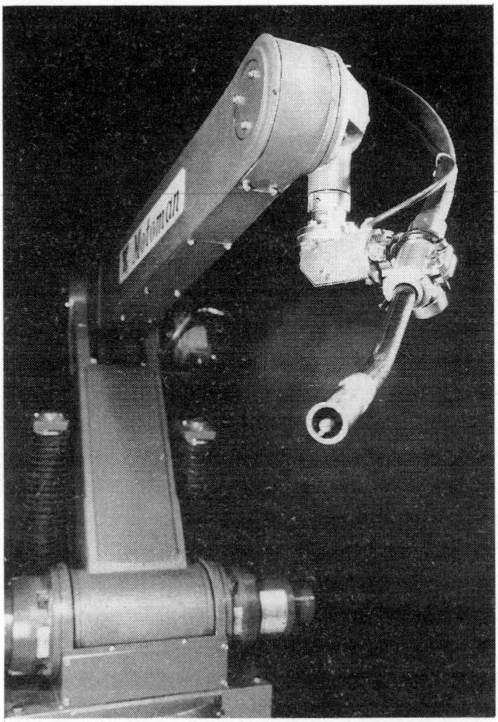

Figure 2.3 Arc-welding robot.

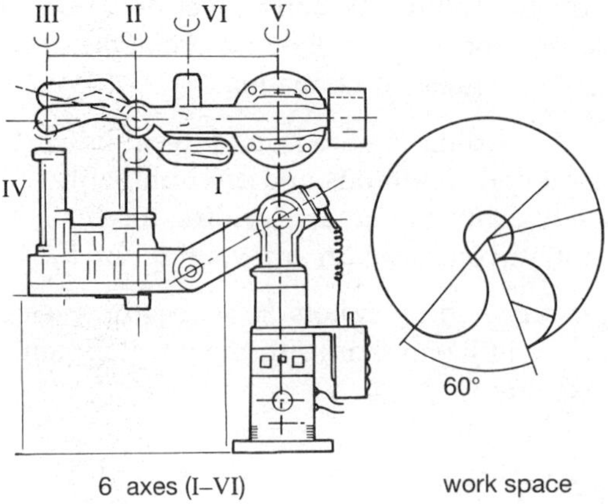


Figure 2.4 Assembly robot with six degrees of freedom: Toshiba TSR-700H.

The conditions for successful assembly robot applications are as follows:

(a) Assembly robot systems must be able to do high-quality work. Applications involving handling, transfer, locating parts and tools, and the grasping and gripping of parts and tools require highly accurate positioning and adaptive speed motion.
(b) Productivity, flexibility and dexterity must be increased.
(c) Industrial robots must be selected such that they do not possess unnecessary functionality, nor should they be too flexible.
(d) Assembly planning and the design of parts and products must be improved so that the work can be performed by a simpler assembly robot.

Until now, most assembly robots in Japan have been of the SCARA (selective compliance assembly robot arm) type (Figure 2.4). Current research and development is being carried out into intelligence, sensors, real-time multi-processor control, and mechanical materials for industrial assembly robots. The results of this are developments in fuzzy and expert intelligent control techniques; vision, tactile, proximity and force sensors; high-speed control systems; quality systems and lightweight materials.

**(c) Deburring robot systems**

Deburring robots, as shown in Figure 2.5, require greater accuracy and dexterity of control, and a greater force control than for other applications. The result of this is that it is difficult to produce a practical

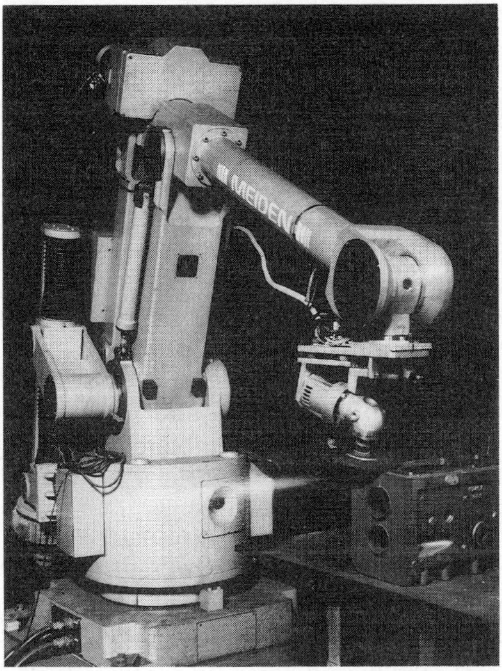


Figure 2.5 Deburring robot.

robot deburring system. Deburring is a necessary finishing process for plastic and aluminium parts and products. Thus a robot deburring system is a necessary component part of a CIM/factory automation system for such applications. Future applications of robot deburring systems would be the finishing of cast metal components.

A number of problems still need to be solved in order to achieve a practical CIM/FA application, the main ones being:

(a) the adaptive control of the motion and force application of a robot's end effector;
(b) the coordination of a robot arm's position with the deburring operation while working between parts and tools;
(c) the development of high-level sensory control systems which use vision, tactile and force sensors;
(d) decision systems for high-level sensory control systems which use high-speed processing algorithms.

### (d) A manufacturing automatically guided vehicle robot

CIM systems require simultaneous control of both information and components as it is important that the information associated with a component be transferred along with the component. To meet this requirement, the manufacturing automatic guided vehicle robot (MAGVR) has been developed in the five-year period from 1984. The MAGVR will be a key element in CIM which will enable an object-oriented manufacturing system to enhance its own functionality in the future.

### 2.3.3 ARCHITECTURE OF A MANUFACTURING ROBOT

A manufacturing robot's architecture can be functionally divided into mechanical systems (mechanisms) and control systems. Mechanical systems are described by machine coordinate systems, and control systems are largely described by task coordinate systems.

A robot has component mechanical structures, namely body, leg, arm, hand, finger, and is known as a manipulating industrial robot (MIR). It is defined by the International Organization for Standardization (ISO, 1988) as a mechanism constructed from three fundamental parts: a primary axis (an arm); a secondary axis (a wrist); and an end effector (fingers and tools). (See Appendix 2.A for the ISO definition of these terms.)

The important functions of an MIR are end-effector motion, velocity and working, and communications. An evaluation of an MIR is dependent upon the control of each of these functions. Motion control is the basic function of an MIR which can be subdivided into joint control and end-effector control. Robots are ultimately evaluated on the effectiveness of their working. However, fine motion and smooth motion are dependent upon the effectiveness of robot control.

A robot control system can be considered to be concerned with the control of working and communications. The control of working can be subdivided into motion control and action control. Figure 2.6 shows a typical robot architecture.

#### (a) Motion control

Motion control includes joint control and end-effector control. Joint control involves specifying angles to describe the rotation of revolute joints, and displacements to describe the sliding motion of prismatic joints. Other joint types are distributed joints (see Appendix 2.A) which can be represented by combinations of revolute and prismatic joints.

End-effector motion control achieves end-effector motion through positional instructions. There are three types of end-effector motion: movement to a position from rest; movement from one position to another as part of a sequence of moves; and continuous path motion. These motion types are achieved using multiple joint position, velocity

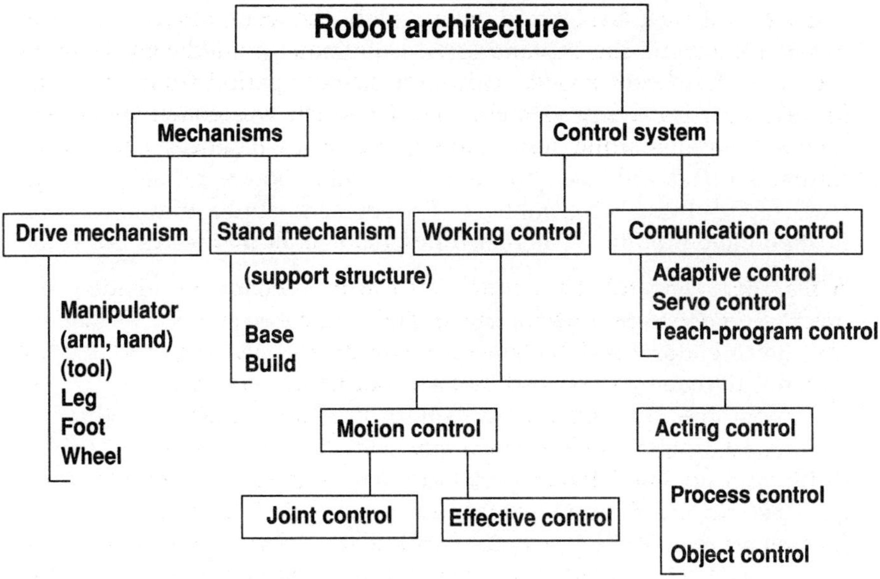


Figure 2.6 Industrial robot architecture. Underlined: mechanical joint motion (machine coordinates). Not underlined: end-effector motion (task coordinates).

and acceleration control. Interpolation is used to define the motion parameters between specified positions.

### (b) Action control

Action control can be subdivided into process control and object control. Process control implements a function in the form of a working sequence for each location of an MIR's program. Object control determines each end-effector position in terms of the speeds and postures of an MIR's body, manipulator and tools.

### (c) Communication control

Communication control comprises sensory control, adaptive control and learning. The control commands are generated by task programming, manual data entry, teach programming, explicit programming and goal-directed programming (see Appendix 2.A).

The basic components of an MIR are its support mechanism and its mechanical system. The support mechanism is responsible for rigidly holding an MIR's base and structure. The mechanical system defines an MIR's coordinate systems. The main coordinate systems are the world coordinates, base coordinates and joint coordinates as defined in Appendix 2.A. The ISO standard also defines the following domains for an MIR: motion space; maximum space; restricted space; operational space; and working space. The operation spaces are where the MIR is controlled while performing programmed motions.

Future MIRs will need to have mechanical joints capable of both mobile and locomotive motion. The present study only considers manipulative motions. Thus an MIR's joint motion is restricted to the following:

(a) an in/out arm traverse (linear motion);
(b) a left/right arm traverse (linear motion);
(c) an up/down arm traverse (linear motion);
(d) a left/right arm turn (rotary motion);
(e) a bending of an arm (rotary motion);
(f) an arm revolution (rotary motion);
(g) a bending of a hand (rotary motion);
(h) a hand revolution (rotary motion).

Single joint motions should be definable; however multiple joint motions are essential for industrial robots working on real applications.

### 2.3.4 DRIVE MECHANISMS

MIRs use electric, hydraulic and pneumatic power for their drive mechanisms. The most appropriate type of drive mechanism is application dependent. In general, a robot capable of lifting large payloads will have heavy-duty actuators. High-speed robots will have lightweight actuators. High-accuracy robots will have specialized multiple actuators.

Each of the major actuator types is appropriate for a typical set of applications:

(a) Electric drive – used for assembly, arc welding, measurement, paint spraying, palletizing, spreading, and inspection tasks which require high flexibility and dexterity.
(b) Hydraulic drive – used for transfer, materials handling, spot welding, assembly, and palletizing operations which have high power requirements and throughput.
(c) Pneumatic drive – used for transfer, loading and unloading, and handling applications which have high speed and throughput requirements.

Existing electrically actuated MIRs are driven by high-performance servo motors which have low moments of inertia and provide high torques. Servo motors, which can be either AC or DC, are very com-

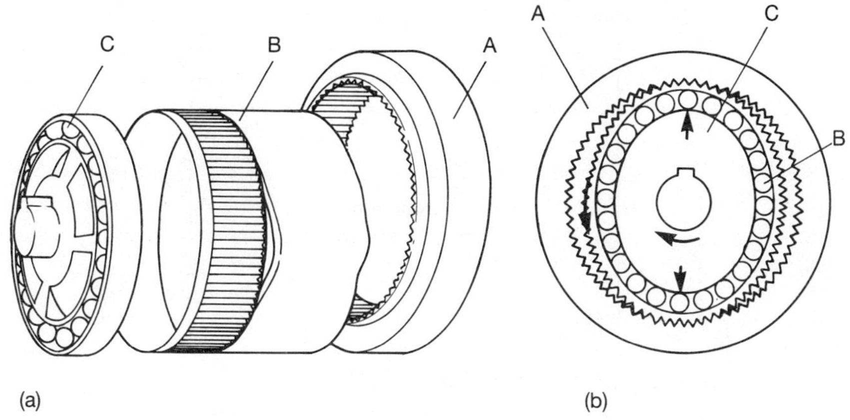


Figure 2.7 (a) Harmonic drive; (b) basic operation. A, circular spline; B, flexspline; C. wave generator.

pact and lightweight. Mechanical convertors can be used to convert rotary motion into linear motion and vice versa. Practical actuators have been developed which use brushless motors, stepper motors and linear motors. Advanced drive mechanisms use direct drive (which obviates the need for transmission systems) and are simple and compact as shown in Figure 2.7.

### 2.3.5 ROBOT MATERIALS

Since the 1950s, materials research related to industrial robots has concentrated on the materials needed to manufacture lightweight, low-cost robots. Several such materials have been developed including an aluminium alloy which is currently being tested for the manufacture of lightweight robots. Alloys which exhibit a quality known as shape memory effect (SME) are also under investigation (Figure 2.8) (Yamodo, 1985).

There are other interesting developments, including a rubber actuator (Kobayashi *et al.*, 1985) which enables a robot arm to be constructed that resembles a human arm and musculature (Figure 2.9). This also incorporates a new actuator which employs metal hydrides with thermal mechanical energy conversion properties. A great deal of Japanese research and development effort into robot mechanisms has concentrated on robot materials. Development projects into new materials are on the increase as the results of such projects renew the currently available robot materials.

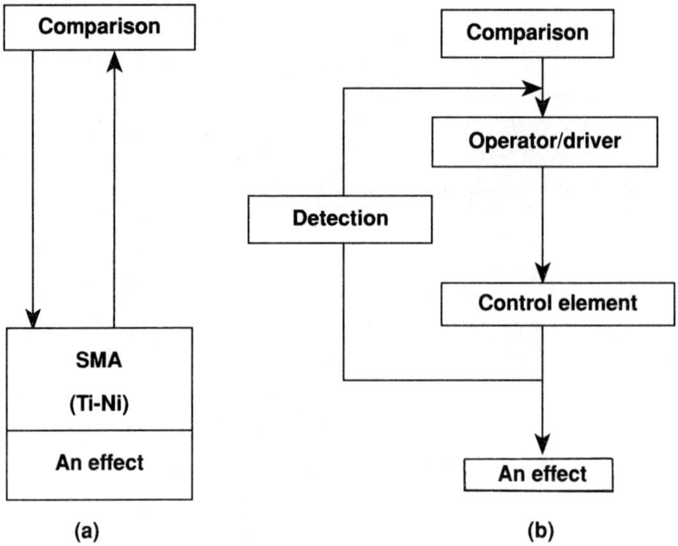


Figure 2.8 (a) Simple SME control system, compared with (b) existing servo control system.

### 2.3.6 COMMUNICATIONS SYSTEMS

Communication is vital for robots in a manufacturing system (Figure 2.10). Developments in manufacturing systems have brought about an increase in robot usage. A consequence of this is the need for complex communication systems with refined hardware and software. Robot communications hardware includes teaching devices (e.g. teach pendants), robot controllers, computers and interfaces. The associated software includes positioning methodologies (e.g. teach and lead-through), control systems, language systems, and simulation systems. This places an emphasis on the importance of robot languages, control systems, simulations, and modelling.

Communications exist between computers and automatic machines, including NC machines, robots and automatic guided vehicles (AGVs). This discussion will only concentrate on communications between computers and robots.

In a similar way to that in which humans require languages to communicate with each other, computers and robots need communications languages. Most robot languages have been developed by manufacturers for a specific robot model, or range of models. This represents a 'tower of Babel' situation, making it difficult to establish communications between different robots.

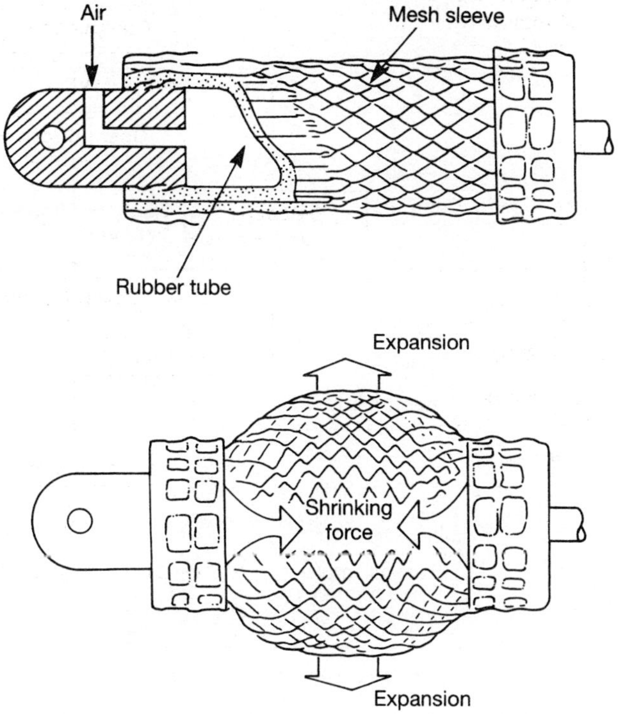


Figure 2.9 Construction of rubber actuator.

Communications systems can be classified according to operations (on-line or off-line), and categories (CAD/CAM) as follows.

Real communications systems can be classified as:

(a) teaching and controlling robots (on-line);
(b) program control of robots (off-line);
(c) communications with robot systems (on-line);
(d) communications with computer systems to control robots (off-line).

Virtual communications systems can be classified as:

(a) designing a virtual robot and robot system (CAD);
(b) inputting a model of a robot or robot system (CAD/CAM);
(c) simulating a robot or robot system (CAD/CAM);
(d) analysing a robot or robot system (CAD/CAM);
(e) fabricating a robot or robot system (CAD/CAM);
(f) programming the operation of a robot or robot system (CAM);
(g) generating a robot program (CAM).

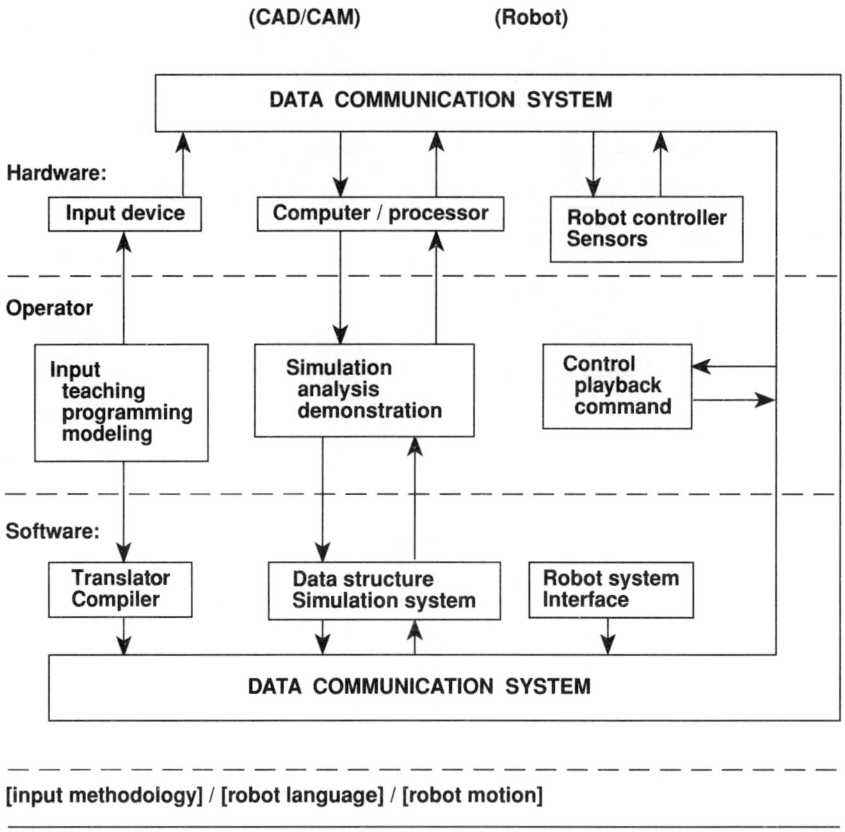


Figure 2.10 Communication system construction and elements.

In real communications systems, an operator controls a robot directly when on-site, and through commands and data when off-site. Operators can control robots in real time; however, this requires robots to have a greater local processing capability than other systems.

Virtual communications systems, being purely software based, cannot control robots in real time. It is also difficult to provide high accuracy and resolution. In fact it is impossible to maintain high performance, high reliability and high adaptability between a virtual communications system and an existing machine (e.g. robot, NC-machine or AGV).

The common data structures and interfaces between systems have an important impact on design. Thus communications systems need to standardize data structures and interface functions. The communications systems which have been described are shown in Figure 2.9.

There are some practical graphical robot virtual communications systems in Japan. A typical system contains a cooperative twin arm simulator, a robot system planner and distributor, an interactive graphical programming and simulation system, and a robot design system.

There are some significant problems in completing the communications and integration of manufacturing systems, in particular the problem of how to input data and commands. One solution is to use planning and control languages. An alternative solution is to use computer graphics and artificial intelligence (AI) systems, both of which show promise.

### 2.3.7 ROBOT LANGUAGES

Technological development in Japan has resulted in numerous applications of robots throughout industry. However, apart from teaching and playback systems, there are comparatively few advanced robot software systems for the SCARA robot configuration.

Most practical robot applications use teach and playback control systems. However, assembly systems using SCARA-type robots use robot languages. These are generally based upon horizontal articulated motions with three degrees of freedom and a compliance control function. SCARA-type assembly robots have two or more axes creating up and down, traverse and rotary manipulating motions.

#### (a) Classification and levels of robot languages

Research and development into robot languages takes place at a global level. Several methods for classifying robot languages have been proposed. These include that of Bonner and Shin (1982) which is based on the level of a programming language. A more recent proposal is that of Inoue (1984) which is based on the presentation of a language's executable functions.

Bonner's classification identifies five levels:

(a) microcomputer level;
(b) point-to-point level;
(c) primitive motion level;
(d) structured programming level;
(e) task-oriented level.

Inoue's classification is as follows:

(a) command level;
(b) primitive motion level;
(c) structured motion level;
(d) object situation level;
(e) task-working level.

**Table 2.2** Existing primitive level robot languages in Japan

| | |
|---|---|
| ARCLE (Assembly Robot Control Language)[a] | Nissan |
| ARL (Assembly Robot Language)[b] | Hitachi |
| AR-Basic | Daikin |
| LUNA (Language for Users Needs and Aims) | SONY |
| MELFA (MEL for FActory robot language) | Mitsubishi Electric |
| MY-Basic | Yaskawa |
| PASLA (Programmable ASsembly robot Language) | NEC |
| PLAW (Programming LAnguage for Welding robot) | Komatsu |
| SCOL (Symbolic COde Language for robot) | Toshiba |
| SERF (Sankyo Easy Robotics Formula) | Sankyoseiki |
| SPEL (Suwa seiko Production Equipment Language)[c] | Seiko Epson |

[a] Suda, 1984
[b] Mohri *et al.*, 1984
[c] Saito, 1984

The current robot programming language classification is defined in ISO (1988), the relevant parts of which are:

(a) manual data-input programming;
(b) teach programming;
(c) explicit programming;
(d) goal-directed programming.

### (b) Practical robot languages in Japan

Recent Japanese research and development into robot languages relates to the primitive motion control level. Command level languages are already implemented in robot controllers. SCARA robot language systems are usually able to process sensory information for existing assembly systems. Primitive motion level robot programming languages used in Japan are listed in Table 2.2.

### (c) Decision-making design and selection

The decision-making processes for robot control using a language are as follows. Adaptive control is used for the robot; thus the communication process is completed through responses, rather than by command. This means that the communication process is completed not by the operator, but by the robot. A consequence of this is that a necessary and sufficient condition for deciding the level of a language is dependent upon the set of available robot functions. This condition $J$ can be expressed in the set form, dependent upon the robot functions $K$.

$$J = \text{solve } (K = K_1 \cap K_2 \cap K_3 \cap \ldots K_{n-1} \cap K_n)$$

where the $K_i$ are robot functional elements.

The control solution $S$ results from the union of prime condition factors $J_n$:

$$S = (J_1 \cup J_2 \cup \ldots \cup J_n)$$

The control execution $E$ becomes the subset of the control solution $S$:

$$E \in S$$

A robot language system is designed or selected using the control solution $S$, a control method.

In general, a language system is dependent upon the communication tools used – in this case a computer system. Both the language expression method and the robot control method must be considered when designing a robot language system. The control of a robot requires procedures for working (with objects). Therefore a robot language $L_r$ is created from a robot's object $O$ and a robot's procedure function $P_f$:

$$L_r \ni (O \cap P_f)$$

A robot language instruction $L_{ri}$ must achieve the total sum of possible practical function expressions, on the control solution $S$. This is because $L_{ri}$ is expressed by the effective cooperation between the action function $Af_{op}$ and the control execution $E_s$:

$$L_{ri} \leftrightarrows \Sigma_o \Sigma_p E_o \cap Af_{op}$$

### 2.3.8 ROBOT INTELLIGENCE

Robots can be considered as typical machines which have artificial intelligence functionality. Robot intelligence functions include knowledge, recognition, communication, understanding, reasoning, learning and decision making. Professor M. Minsky initiated research and development into artificial intelligence in his discussion 'Intelligent machinery (Turing, A.M.)' in 'Steps towards artificial intelligence (MIT)'. Japanese investigations began in 1979 with the fifth-generation computer system (FGCS) project. This resulted in the foundation of the Institute for New Generation Computer Technology (ICOT) in 1982.

Japanese research and development projects into artificial intelligence (AI) have resulted in the personal sequential inference engine (PSIM) and the sequential inference machine programming and operating system (SIMPOS). A major research and development project has recently been completed which plans the creation of adaptive robots for work in the ocean, nuclear power, underwater, firefighting and disaster prevention environments. A new research and development project is being planned in AI for the intelligent manufacturing

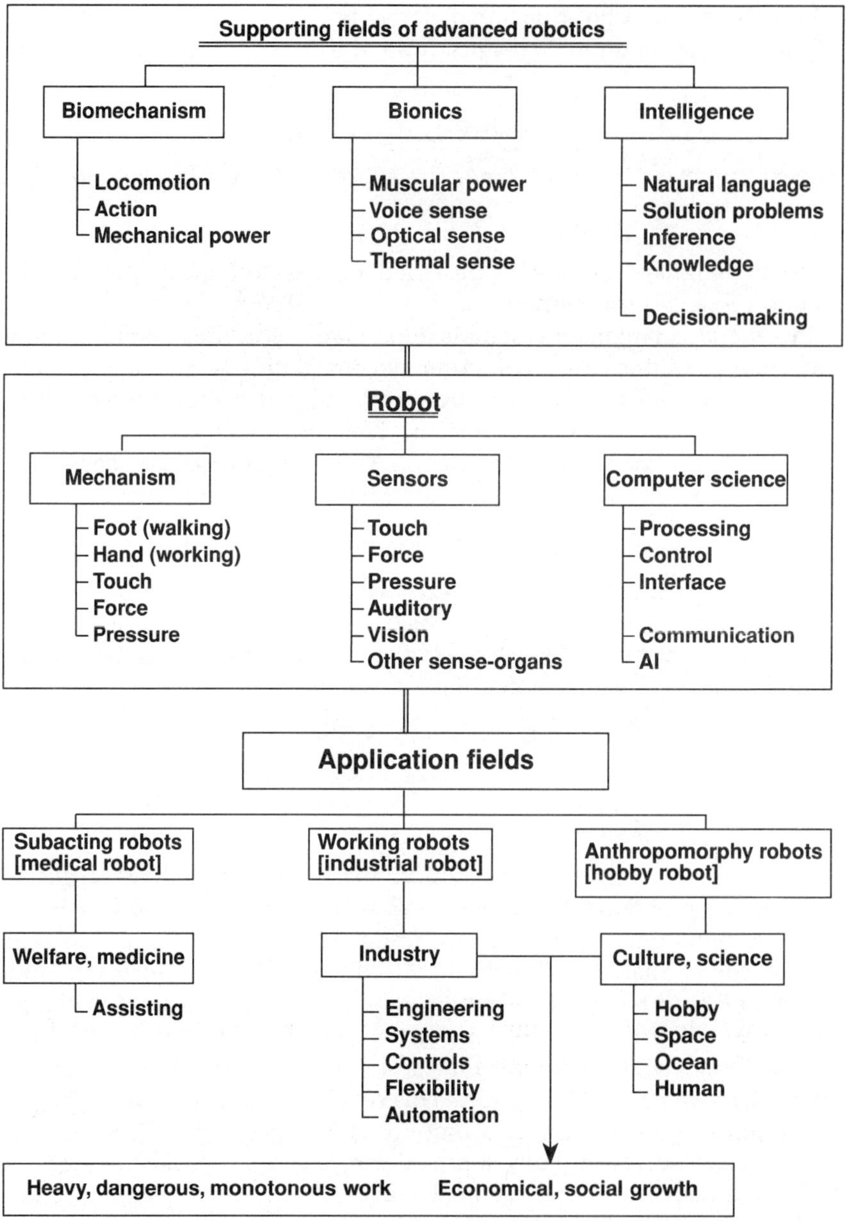


Figure 2.11 Robotics and application fields.

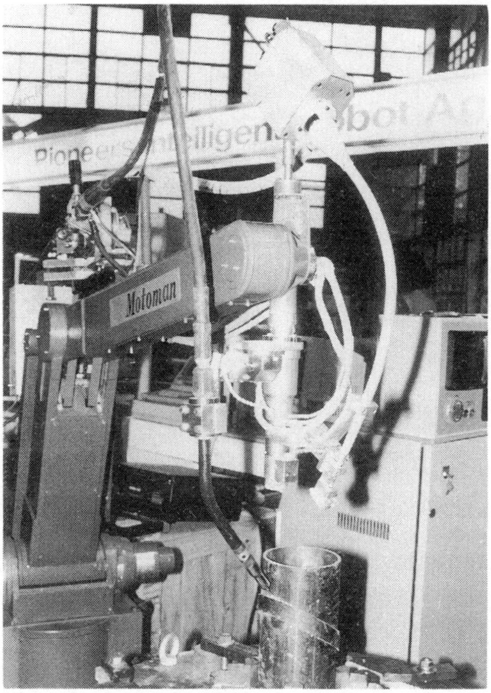


Figure 2.12 Robot hand with vision sensor.

system (IMS). This Japanese proposal will be an international venture involving Europe, the USA and others.

Existing intelligent adaptive control robot systems include the following practical systems (Figure 2.11):

(a) *Seam-tracking robot system using a molten pool vision sensor*: This arc-welding robot, the Motoman-L10w, has a hand-assembled sensor and torch (Figure 2.12). A fibre-optic scope enables the detection of molten pools during arc-welding. A laser is used to locate the seam and to detect the welding torch. The fibre-optic scope and the laser can be rotated independently. The manipulator has a rolling motion using a motor and gear mechanism. The two axes of the optical sensor are coincident with the molten pool.

(b) *Deburring robot system using a force sensing tool*: An advanced robot, the MeiRobo GWR60, has been developed to meet the problems of

Figure 2.13 Deburring tool with force sensor.

economy, quality and functionality in industry (Figure 2.13). This robot uses a force control system to move its hand by sensing load variations.

(c) *Assembly robot system using two cooperating arms*: The FAROT-M6 250II robot has two articulated, six degrees-of-freedom arms (Figure 2.14). One arm can supply, transfer and position a part or product. The other arm can insert one part into another. This enables the robot to use its two arms to process a product completely.

### 2.3.9 STANDARDIZATION OF INDUSTRIAL ROBOTS IN JAPAN

Today's Japanese suppliers conform to the ISO/IEC standards which are accepted by both manufacturers and users. Although most adhere to ISO/TC184/SC2, the Japanese Industrial Robot Association (JIRA) has outlines modifications to the ISO standards as listed in Appendix 2.B.

ISO/TC184 *Industrial Automation Systems*/SC2 *Industrial robots* was established in 1984. TC184 was restructured in 1984, however SC2 was not changed at all. SC2 now has six working groups which are as follows:

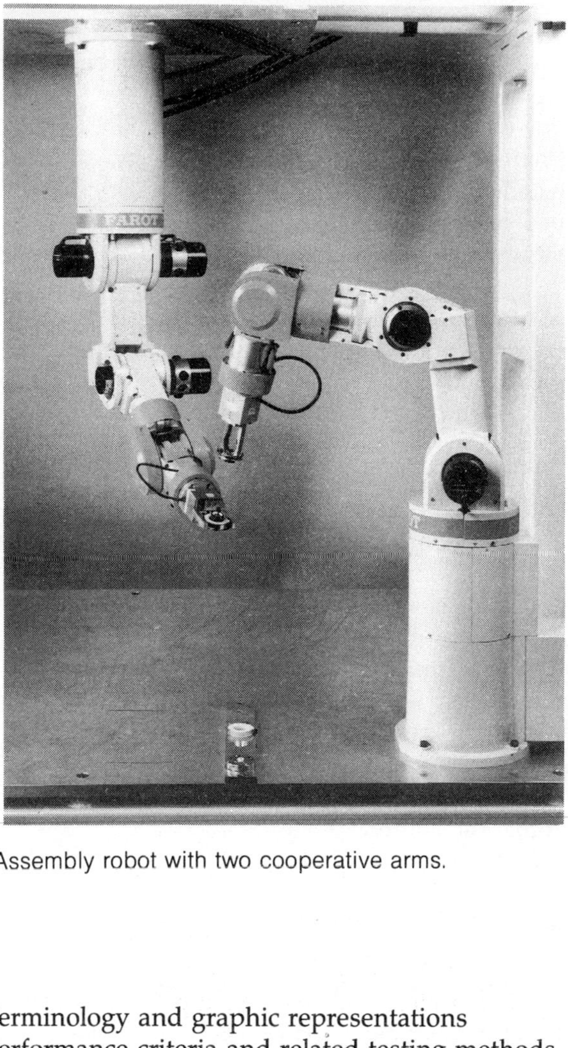

Figure 2.14 Assembly robot with two cooperative arms.

(a) WG1: Terminology and graphic representations
(b) WG2: Performance criteria and related testing methods
(c) WG3: Safety
(d) WG4: Programming methods and languages for MIR
(e) WG5: Mechanical interfaces
(f) WG6: Robot companion standard to MMS.

The standards listed in Appendix 2.B and related standards were established by ISO and the International Electrotechnical Commission (IEC). The industrial robot standardization of ISO/TC184/SC2 required liaison with other ISO committees and sub-committees, and with IEC. It also required an awareness of Open Systems Interconnection (OSI) activity in information communication and network communication standards.

### 2.3.10 ADVANCED ROBOTS

Until recently, research and development in robotics concentrated mainly on the areas of high power, high speed and high accuracy. This focus has now changed to adaptability, operability and reliability. The need for human–machine interfaces between operators and robots has been acknowledged. This includes the relationship between operators and machines in human–machine systems, CAD/CAM for systems engineers designing robot systems, FA, FMS and CIM (Takashima, 1982).

The future of robotics research and development rests on communications, sensory and working capabilities, and if possible reasoning and creative capabilities. Some applications will require robots to move freely, at will, around factories with established advanced manufacturing systems. These increasingly sophisticated demands for advanced robots make it unlikely that their mass production will be possible. It has been said that the prosperity of the Japanese robot industry is based upon the national characteristics of diligence and adaptability. Japanese workers in industry have a co-existence with tools and machines, and have shared in the resulting prosperity.

One of the many problems associated with robots and robot systems in Japan is the small number of practical applications provided by the universities, research institutes, government laboratories, and other organizations which conduct research and development in robotics. This problem has been addressed by the establishment of the National R&D Project on Advanced Robotics and the International Robotics and Factory Automation Centre.

In particular the Japanese robot manufacturing industry is conducting research and development in such areas as zero maintenance. In the current Japanese market, industrial robots are classified by the Japanese Industrial Standards (JIS) in cooperation with the Japanese Industrial Robot Association (JIRA), as follows (JIS, 1986):

(a) General classification:
    (i) operating robot
    (ii) sequence control robot
    (iii) playback robot
    (iv) numerically controlled robot
    (v) intelligent robot
    (vi) sensory controlled robot
    (vii) adaptively controlled robot
    (viii) learning controlled robot.
(b) Classification by control method:
    (i) servo controlled robot
    (ii) non-servo controlled robot

   (iii) continuous path (CP) controlled robot
   (iv) point-to-point (PTP) controlled robot.
- (c) Classification by motion:
   (i) cartesian coordinate robot
   (ii) cylindrical polar coordinate robot
   (iii) spherical polar coordinate robot
   (iv) articulated robot.

Possible future applications include robots with 'seeing' eyes, and touch sensitive skin (Kitagawa, 1984).

## 2.4 NUMERICAL CONTROL ENGINEERING AND FACTORY AUTOMATION

Research and development is being carried out on a computer numerically controlled (CNC) system which is to be used as a simulator and as a learning system.

### 2.4.1 AN ADVANCED NC SYSTEM

A major challenge is to produce an intelligent and customizable system which is capable of representing the engineering and information specifications for manufacturing machines such as NC machines, robots and installation systems. This must accommodate the need for standardization in industry, classification, characteristic architecture, communications, and software in CIM and future manufacturing systems.

Research is being conducted into engineering activities for older skilled operators. Various computer-assisted systems related to manufacture are also being developed in Japan. The 'communication mature programming system for engineers' (COMPOSER) is a CNC system equipped with a programming simulator and an intelligent training system. COMPOSER is designed to improve both an operator's efficiency and environment.

### 2.4.2 COMPOSER COMMUNICATIONS

COMPOSER's communications concept is based on the concept of an interface between a mature, experienced operator and a CNC machine. It is expert-system based and uses the following components for its knowledge-base requirements:

- (a) a communication system to increase the speed of CNC data input;
- (b) system functions to make the best use of a mature engineer's knowledge (know-how) and skill (techniques);
- (c) a manufacturing database which is built into the system.

The following functions are under development:

(a) simplified NC programming systems using both a descriptive language and computer graphics;
(b) improved programming input/output devices;
(c) practical NC tool path simulation;
(d) an expert system facilitating the learning and referencing of experience.

CNC programming systems which use computer graphics are popular communication tools in total CAD/CAM systems. Existing systems pose problems in the areas of safety, useability and productivity for mature, experienced engineers and operators. The COMPOSER system addresses these problem areas.

### 2.4.3 CONSTRUCTION OF ADVANCED CAD/CAM SYSTEMS

These communications systems use both non-descriptive graphics and descriptive programming language methods.

The CAD component of COMPOSER can determine a product's parameters, including its functions, shape, specification, and process planning aspects. The CAM component can schedule manufacturing, simulate a cutting path checking for interference between tools, parts and environment, and make decisions regarding tools and cutting conditions as follows:

(a) A common database can use the graphics and language methods in tandem, employing the shape's parametric and geometric elements.
(b) The graphics and language methods can interchange pictures and languages, making corrections, insertions, deletions and additions to the common database in real time.
(c) Graphics enhance human perception.
(d) The language method features algorithmic extensions.
(e) The knowledge database and processor assist engineers to design shapes they manufacture using standard technical information and accumulated data.

### 2.4.4 THE CONCEPT OF COMPOSER-I

Japanese research and development in manufacturing engineering and systems is generally aimed at the development of advanced and mature engineering. Unlike the past goals of high power, high speed and high accuracy, advanced and mature engineering concentrates on the interface between the operators and the machines of a manufacturing system. This interface needs to operate at a higher level than the usual human–machine system relationship. System designers need to become aware that advanced systems, particularly CAD/CAM, FA, FMS and CIM, require a much more sophisticated human–computer interface than is currently offered. Current machine-shop design work concentrates on adaptability, operability, reliability and safety. The system design of COMPOSER is founded on encouraging mature workers to use a CAD/CAM system actively, and to provide a safe working environment.

It is a major objective that the prototype COMPOSER-I system be extensively used by mature engineers both as a learning and training system, and for practical applications. To facilitate this, the core programming components of COMPOSER-I are a translator and an editor which enable mature engineers to learn and train themselves.

Macros have been implemented, using heuristics, which enable mature engineers to implement tasks on the system using simple specifications methods. This enables them to experience the creation of tool paths and the determination of machine conditions. These macros are representative of the use of programming languages and graphical operations for performing practical tasks.

The relationship between mature engineers and operators is that of practical assistance, cooperation, and the improvement of the COMPOSER-I environment in these areas:

(a) Practical assistance:
  (i) knowledge system development;
  (ii) know-how on machining metals (modelling a removal procedure and process);
  (iii) data files on cutting;
  (iv) jig design (decisions about materials);
  (v) learning from an experience-based knowledge system.
(b) Cooperation: the development of an interactive programming system (cooperation of programming tools, producing NC programs, and assistance in operation and data verification).
(c) Environment improvement:
  (i) development of I/O devices (tools);
  (ii) simplification of loading, unloading and setting of parts and materials;
  (iii) troubleshooting and maintenance;
  (iv) improved working environment and cleaner working place.

### 2.4.5 COMPOSER-I EVALUATION

There are a number of system products in Japan which are concerned with NC and CNC programming; these have mainly been established in manufacturing engineering departments. Some of these systems are simplified interactive CNC programming systems. It has been found that, in general, these systems are not considered useful by mature engineers and operators. COMPOSER-I has been tested by mature engineers with good results. The system is not yet complete; it needs further research and development.

### 2.4.6 FINISHING OF MACHINE PARTS FOR ASSEMBLY

Research and development into the finishing of machine parts covers a number of areas including: polishing; grinding; brushing; blasting; electrolytic machining; electrochemical machining; electron beam machining; and skilful cutting for deburring, edging (chamfering and finishing), and surface conditioning. New deburring and surface conditioning methods use electrolytic grinding, electrolytic polishing, and electron beam cutting. These processes must be carried out before assembly can proceed, hence there is much activity in these areas.

Advanced manufacturing and CIM systems require practical methods of incorporating automatic deburring. Japanese research and development work in automatic deburring concentrates on industrial robot systems. This is in response to the demands of both manufacturers and workers for a safe, automatic factory environment. There are however many problems which put the progress of automatic deburring behind that of other manufacturing techniques. The reason for this is that burr formation is both complex and irregular. It is dependent upon the material's properties, tools, and the machining methods, procedures and conditions. Many of these dependent factors generate burrs with characteristic shapes and sizes. Thus the deburring method can be selected according to the type of burr which has been generated.

Deburring operations are carried out in noisy and dirty environments which have a detrimental effect on workers' health, hence the development of a robot deburring system is considered a priority. Most deburring machines currently in operation are manually operated due to the lack of automated systems.

The Japanese government organizations and industry have carried out tests robot deburring systems which have led to the development of practical deburring tools and robot systems for automatic processes in FMC, FMS and factory automation. They have not yet been applied to iron casting and moulding processes, but have mainly been used in factories deburring plastics and aluminium parts. Japanese factories often do not have enough deburring systems to satisfy the throughput requirements of assembly and transfer systems.

Automatic machinery which takes product and part designs, and processes raw materials through to finished parts, require automated post manufacturing processes. These include finishing, assembly, painting/spraying, testing, packaging and product delivery. It is anticipated that these processes will be integrated into fully automatic systems. This integration poses the following fundamental problems to Japanese manufacturing, particularly regarding the use of robot systems.

Each manufacturing cell should be completely self-contained. This leads to the problem that when parts are being transferred from one cell to another, an unreasonable problem can be imposed upon cells receiving them. A part which retains burrs from a machining cell will cause problems in an assembly cell. The major difficulty of integration is that of communicating information and providing processing information for each part at each cell. Thus an essential component of an advanced manufacturing system is a cell-defined communications system such as a LAN or WAN in a CIM system.

## 2.5 CIM SYSTEMS AND MAP/OSI IN JAPAN

Factory automation has met recent demands for improved productivity. It has shown that NC machines, MIRs and other mechatronic machines are of high quality and can be used with flexibility and reliability in a manufacturing environment. These high-quality manufacturing systems require that each manufacturing cell is responsible for automating a particular aspect of the manufacturing process. Each cell requires a software system which has been developed as part of a CAD, CAM, CAE, CAT or other computer-assisted system.

Manufacturing systems were first evaluated in their role of improving productivity in the mass-production automated assembly lines of the automobile industry. Next, they were evaluated for providing flexibility of batch production in the electrical and electronic industries. The latest evaluation is of the CIM system using network communication and the Manufacturing Automation Protocol (MAP) for the industrial automation of all types of factory.

Future manufacturing systems will be dependent upon computer networks because they are complex, decentralized, information dependent and intelligent. The autonomous distributed/decentralized control-integrated manufacturing system (ADCIMS) describes the ideal situation, but it does not yet exist. This ideal ADCIMS will require advanced manufacturing cells equipped with industrial robot systems, NC machines and other manufacturing systems incorporating intelligence, communications and other special functions. It will also require complete CAD, CAM, CAE, CAT, CNC and DNC technologies and systems, which were in practical use in the 1980s as tools of automation.

Ideal manufacturing systems have several levels of operation. The

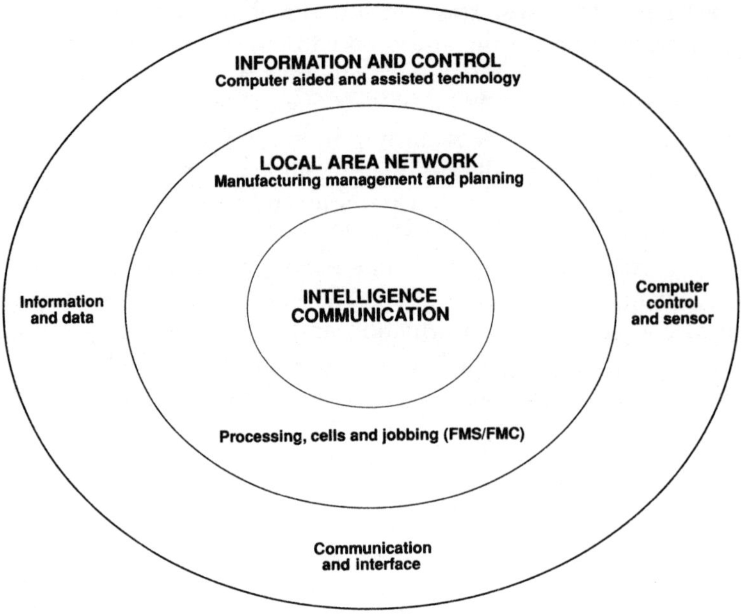


**Figure 2.15** A CIM/IMM system reference model.

top level is the ADCIM which incorporates the highest level of intelligence and technology. The second level, which may become a practical system in the future, is the intelligent management and manufacturing (IMM) system (Figure 2.15). The third level is CIM/FA which is made up of materials handling systems, machining cells, assembly cells, testing cells and packaging cells. Each process transmits and receives the data associated with parts and products as required. This requires a human interface to CIM/FA systems as well as interfaces between systems.

A standardized communication protocol for CIM/FA network system interfaces is MAP 3.0, which the World Federation of MAP/TOP User Groups (WFMUG) has frozen for several years since 1988.

The fourth level of an ideal manufacturing system is the factory automation system which generally uses RS232C, RS422 or other such interface with non-protocol communication or simple protocol communication. These communications methods are unsuitable for a CIM/FA network.

### 2.5.1 FLEXIBLE AUTOMATION MANUFACTURING SYSTEMS CASE STUDIES

There have been numerous studies and projects on manufacturing systems throughout the world. These include:

(a) the US Air Force's integrated computer-aided manufacturing (ICAM) programme;
(b) DRAPA;
(c) automated manufacture research facilities (AMRF) in the USA;
(d) the National Institute of Standards and Technology (NIST) studies of the ESPRIT programme on computer-integrated manufacturing, open systems architecture in Europe;
(e) the Japanese Technology Association's methodology for unmanned manufacturing (MUM);
(f) the IROFA studies of FMSC and IMS.

Manufacturing research and development in the USA and Europe, which began with basic functional concepts of a computer system, is moving on to the applications of computers to factory design. Examples of this are the Society of Mechanical Engineers' CIM-Wheel, and the CAM-I discrete parts manufacturing model (DPMM). Japanese studies are moving from factory technologies to the application of computers to automation systems. The first manufacturing concept is MUM. Studies are also being conducted into CIM and IMS design with artificial intelligence, office automation and factory automation.

The production of CIM systems which apply artificial intelligence, office automation, factory automation, information technology and standardization technologies are of common interest in Japan, Europe and the USA. Advanced practical technologies exist in the fields of biomechanics, manufacturing, expert systems which incorporate databases, and other manufacturing systems.

The final manufacturing system is CIM. This presents many structural problems with practical database systems using real data and information, and with communication systems which network stations, machines and humans.

The interfaces between operators and machines are important. Studies in this area include human integrated manufacturing (HIM) of the type described in section 2.4. When structuring CIM systems, it is necessary to integrate HIM and CIM so as to build the essential relationship between human beings and machine elements. Studies in Japan have centred on the usual manufacturing technologies of design, machinery, assembly, testing and control. Information technology related to manufacturing has so far received little attention.

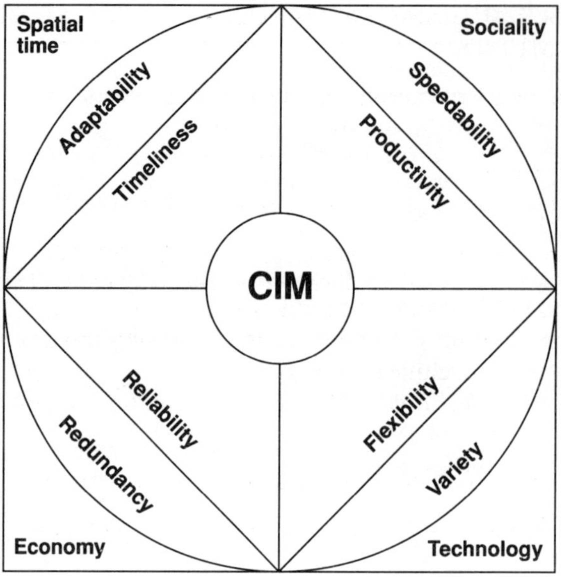


Figure 2.16 CIM integrated and structured by various parameters and variables.

### 2.5.2 COMPUTER-INTEGRATED MANUFACTURING SYSTEMS

It has been mentioned previously that current automated manufacturing systems tend to be aimed at adaptability, redundancy, variety and speed. The four elements of productivity, flexibility, reliability and timeliness are requisites in manufacturing automation systems. They are also just as necessary for advanced manufacturing systems, such as FMS and FA, which are necessary for the satisfactory achievement of social, economic, technical and time-scale demands. These relationships affect the way in which CIM is integrated and structured (Figure 2.16).

### 2.5.3 CURRENT CIM IN JAPAN

Many CIM systems have been developed in Japan, both for management communication and at the manufacturing automation level.

There are several methods of classifying CIM in Japan, due to lack of agreement. Typical considerations are CIM integration of management information databases, manufacturing processing methods, manufacturing systems, materials and resources, products, machines and communications. Table 2.3 shows a simple classification based on spaces and effects. These spaces are established with human interfaces and system interfaces. The human interface in CIM is particularly important

**Table 2.3** JIRA adaptations of ISO standards

| |
|---|
| The Japan Industrial Standards (JIS) |
| Glossary of terms for industrial robots (JIS B 0134-1986) |
| Symbols for industrial robots (JIS B 0138-1980) |
| Standard form for indicating characteristics and functions of industrial robots (JIS B 8431-1988) |
| Measuring methods for characteristics and functions of industrial robots (JIS B 8432-1983) |
| General code for safety of industrial robots (JIS B 8433-1986) |
| Identification symbols and colours for the operator controls of industrial robots (JIS B 8434-1984) |
| General code of design for modularization of industrial robots (JIS B 8435-1986) |
| Circular flanged mechanical interfaces of industrial robots (JIS B 8436-1989) |
| Japan Industrial Robot Association Standards (JIRAS) |
| Standardization of electric signals of industrial robots (JIRAS 1006-1986) |
| Standardization of piping between machine and industrial robots (JIRAS 1008-1986) |
| Standardization of wiring between machine and industrial robot controllers (JIRAS 1009-1986) |
| Standard interface between machine and industrial robot controllers (JIRAS 1010-1986) |
| Holders of industrial robots |
| Current standardization |
| Standardization of robot language (ICR/high-level language) |
| Robot message system |

because a completely unmanned manufacturing system is likely early in the 21st century.

Current Japanese technological studies are concentrating on communications in fuzzy processing and neural networks. Japanese manufacturing research and development in CIM is directed towards intelligent systems.

All CIM systems have information systems which distribute both executive commands and information on parts, products, tools, machines and systems. This is achieved through communication systems such as local and wide area networks, existing field buses, and communications interfaces (RS232C, RS422, HDLC and Ethernet). Figure 2.17 shows a typical example of a recent CIM/factory automation system in Japan.

### 2.5.4 PRACTICAL MANUFACTURING SYSTEMS

Existing CIM, CIM/FA and CIM/IMM systems are used particularly on local area networks, including the manufacturing message specification

Figure 2.17 A typical recent CIM/FA system in Japan.

(MMS). Japanese factories need to install multiple communications systems, as they are getting confused by complex processing and connections in manufacturing systems. Japanese manufacturing systems will attempt to use MAP and Technical Office Protocol (TOP) with MMS in a manufacturing local area network.

It is expected that MMS application layer protocols will be used in future CIM systems. The first stage of communication is between station and station. The second is between cell and cell. The third is between machine and machine on the shop floor. The fourth is between stations, cells and machines. The types of cell include distributing cells, machinery cells, assembly cells, testing and measurement cells, and transfer cells. These communicate information and data within factories over the MAP and TOP manufacturing networks.

The Japanese Factory Automation Interconnection System (FAIS) is similar to a minimal MAP and is becoming a practical standard. A MAP test centre has been established in Japan and is described opposite.

## 2.6 MANUFACTURING COMMUNICATION STANDARDS ENVIRONMENT IN JAPAN

Centralized CIM is realized with a communication system that integrates in-company information processing systems and provides the necessary information at the right time.

A LAN, which operates at high speed and permits flexible communication around the system connected to it, is expected to serve as an optimum inter-house network system. It also permits the formation of partially unified systems (the so-called 'islands of automation'), as the protocols for communicating between systems have not been unified. The existence of many different LAN protocols means that an interface is required for each system connected to the LAN. This also makes communication between systems difficult. Hence there is a crucial need for a LAN based on internationally accepted communication rules, to facilitate the development of CIM systems.

### 2.6.1 INTERNATIONAL COMMUNICATION STANDARDS OSI/MAP

OSI is a standard communications protocol for worldwide networks. The ultimate aim of OSI is to provide economical and efficient communication in many different systems. In order to apply OSI to a factory LAN, the MAP of each of the seven OSI layers defines the profile for manufacturing environments. MAP can connect through wide area networks (WAN) to other domestic factories and even overseas sites. The result is a huge international manufacturing capability.

### 2.6.2 MAP TEST CENTRE

Products based on MAP specifications must be tested to ensure that they conform to the specification. If they are tested according to different criteria and by different test methods, they may not be able to communicate. To ensure compatibility, MAP products should be tested for conformance by a third party. The conformance test cannot guarantee complete compatibility, but it can at least guarantee complete interconnectibility. The world's first MAP TEST CENTRE (MTC) was established for this purpose in 1989.

#### (a) Test of MAP products

MAP is classified into full MAP, which is based on the OSI reference model, and mini-MAP (including the FAIS specification) which is based on the first, second and seventh layers of the OSI reference model. The OSI model defines each of the seven layers in detail. The testing proceeds from the lower to the higher levels. MAP equipment passing the tests is recognized as being highly reliable.

### (b) MAP conformance testing

Conformance with MAP product specifications must be tested in an identical test environment before the product is delivered to the nearest market. MAP products with different interpretations of the specifications, or with undetected programming errors, can disturb interconnections with other MAP products.

### (c) International harmonization

MTC is equipped with systems for performing full-MAP and mini-MAP tests based on MAP specification version 3.0. This includes the IEEE 802.4 physical layer and MAC sub-layer of the data link layer. The conformance test systems being used in MTC were developed by the Standards Promotion and Application Group (SPAG), the Industrial Technology Institute (ITI), and the Corporation for Open Systems (COS). These are internationally recognized as reliable organizations for developing MAP, TOP and OSI conformance testing systems. MTC must harmonize internationally with other test facilities with the co-operation between SPAG, COS, IROFA, the Interoperability Technology Association for Information Processing (INTAP), and other OSI, MAP and TOP standards organizations. MTC will contribute to the popularization of OSI and MAP worldwide.

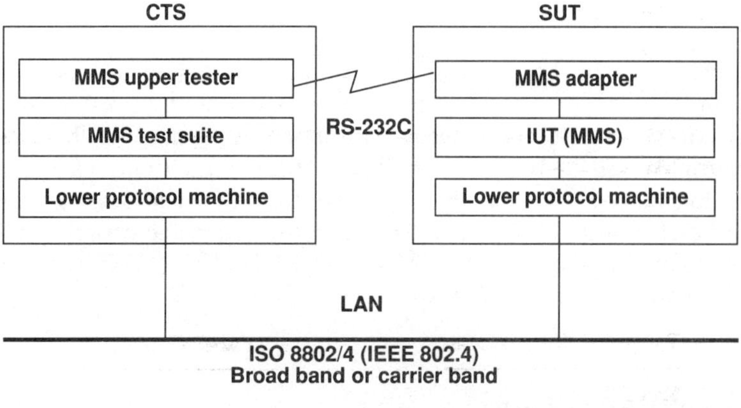


Figure 2.18 Conformance test system (CTS): typical configuration. SUT, system under test; IUT, implementation under test; MMS, manufacturing message specification. The upper tester, located higher than the IUT in the SUT, observes and controls the IUT. The test suite is a complete set of test cases, possibly combined into nested test groups, that is necessary to perform conformance testing for an IUT. Conformance testing requires an RS-232C to exchange control information between the CTS and the SUT, in addition to LAN communication.

**(d) Conformance test system**

The physical layer test system, the MAC sub-layer test system and other protocol test systems for OSI, from the seventh layer down to the third layer and the logical link control sub-layer, are established in MTC (the MAC sub-layer and the logical link control sub-layer form the second layer, the data link layer, of the OSI model). An example of an MMS conformance test system is shown in Figure 2.18.

**(e) Test service**

Conformance testing is carried out on each protocol in Table 2.4 by conformance test flow (Figure 2.19). The necessary documents for MAP conformance testing are:

(a) application for MAP conformance testing
(b) MAP conformance testing agreement
(c) system conformance statement (SCS)
(d) protocol implementation conformance statement (PICS)
(e) protocol implementation extra information for testing (PIXIT)
(f) specification for the upper tester
(g) client manuals
(h) instructions for handling the system.

The following test reports on MAP products passing the conformance test are issued by MTC:

(i) system conformance test report (SCTR)
(j) protocol conformance test report (PCTR).

These reports are necessary for clients when applying for the certification of MAP products.

Table 2.4 Simple classification of spaces and effects

| Space | Effects |
|---|---|
| Human being | Technical tradition |
| Cooperation | Machine's assisted work |
| Program | Reduced working members |
| Automation | Systematic |
| Multi-complex | Flexibility |
| Integrated | Adaptability |
| Communication | Real time (just in time) |

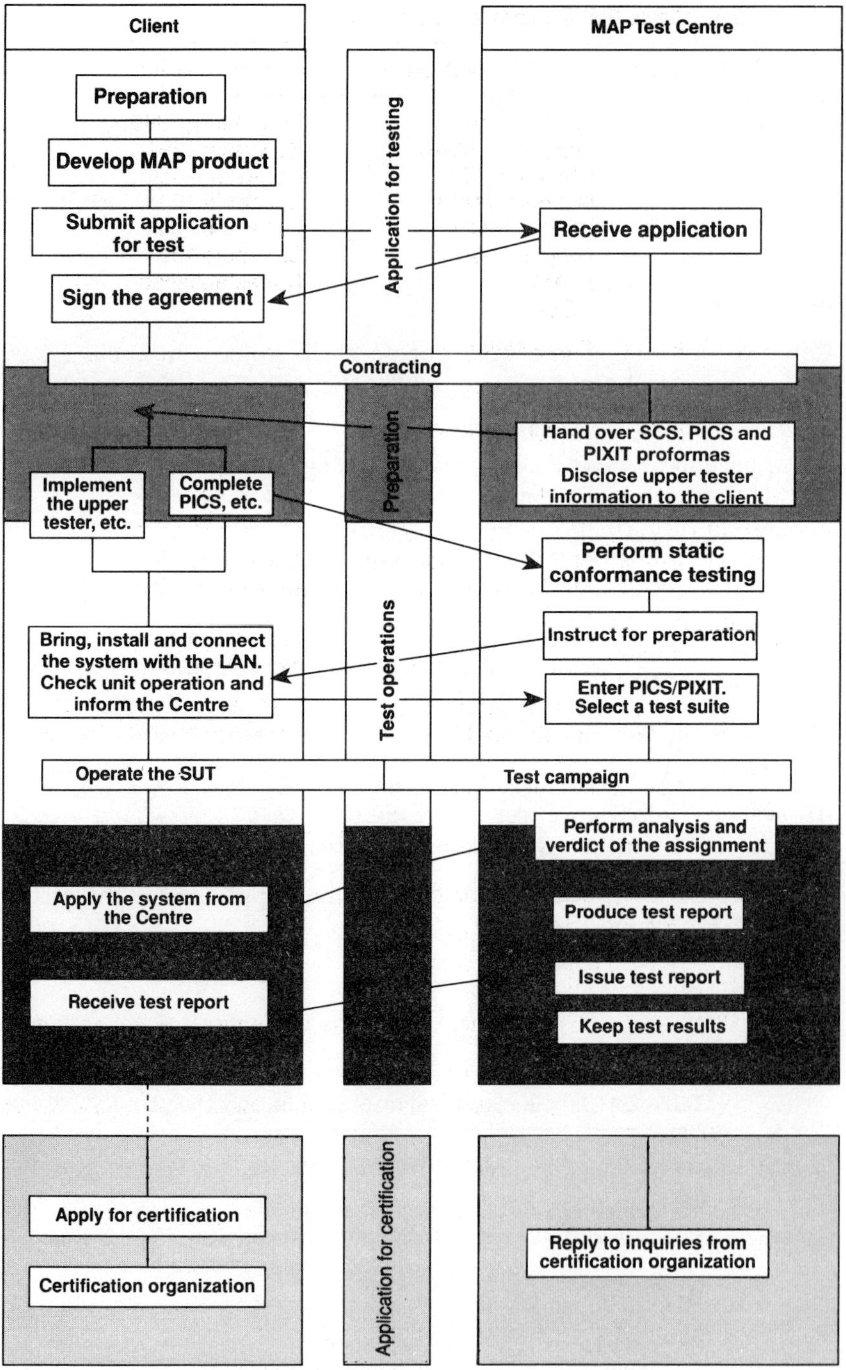


Figure 2.19 Conformance test flow.

**Table 2.5** Conformance test list

| Layer | IUT (protocol) | Service start-up |
|---|---|---|
| 7 | MMS (full MAP) | End-March 1990 |
| | MMS (mini MAP) | Schedule 1991 |
| | NM (full MAP) | Not established |
| | NM (mini MAP) | Schedule 1991 |
| | DS (full MAP) | Not established |
| | DS (mini MAP) | Schedule 1991 |
| | FTAM | Not established |
| | ACSE | * |
| 6 | Presentation | * |
| 5 | Session | * |
| 4 | Transport class 4 | Not established |
| 3 | ES-IS | Not established |
| | Router | Not established |
| | IN | Not established |
| 2 | LLC Type 1 and Type 3 | Schedule 1991 |
| | MAC carrier band | End-March 1990 |
| | MAC broad band | End-March 1990 |
| | MAC bridge | Not established |
| | MAC immediate response | Not established |
| 1 | Carrier band | End-March 1990 |
| | Broad band | End-March 1990 |

* Cannot be tested independently

## REFERENCES


Bonnor, A. and Shin, G. K. (1982) A comparative study of robot languages. *IEEE Computer*, December, 82–96.

FGCS '84 *ICOT, November 1984.*

Inoue, H. (1984) Towards advanced robot programming. *Journal of the Robotics Society of Japan*, April, 3–6.

ISO Technical Committee (1988) Sub-committee 2, Technical Report TR83730.

Kitagawa, T. (1988) Computer science and technologies. *Japan Annual Reviews in Electronics*.

Kitagawa, T. (1984) Intelligent robots and their applications. *Computers and Telecommunications*, 189–275.

Kobayashi, S. *et al.* (1985) Artist robots. *Robot*, No. 47, April, 43–49.

Mohri, S. *et al.* (1984) ARL (Assembly Robot Language). *Journal of the Robotics Society of Japan*, December.

Saito, K. (1984) SPEL (Suwa Seiko Production Equivalent Language with CAD). *Journal of the Robotics Society of Japan*, December.

Suda, H. (1984) ARCLE (Assembly Robot Control language). *Journal of the Robotics Society of Japan*, December.

Takashima, S. (1982) CAD system for CAM assisted with robots. *Operations Research*, No. 7, September, 511–525.
Yamamoto, Y. (1985) Development of arm with five fingers, an application of SMA to robotics. *Robot*, No. 46, January, 81–86.

## APPENDIX 2.A GLOSSARY OF ISO DEFINITIONS

**Adaptive control** A control scheme that adjusts the control system parameters from conditions detected during the process.
(ISO TR8373: 6.3.4)

**Arm (primary axis)** An interconnected set of links and powered joints comprising longitudinally shaped members which support, position and orientate the wrist and/or an end-effector.
(ISO TR8373: 4.2)

**Base coordinate system** A coordinate system referenced to the base of the robot.
(ISO TR8373: 5.4.2)

**End-effector** A device specifically designed for attachment to the mechanical interface to enable a robot to perform its task (e.g. gripper, welding gun, spray gun).
(ISO TR8373: 4.9)

**Explicit programming** A programming method in which end-effector positions or the desired path are explicitly defined.
(ISO TR8373: 6.2.4)

**Goal-directed programming** A programming method in which the task is defined but the actual end-effector path required to achieve the task is not prescribed.
(ISO TR8373: 6.2.5)

**Joint coordinate system** A coordinate system referenced to a robot's joints. Each joint coordinate system is defined relative to the preceding joints, or to some other coordinate system.
(ISO TR8373: 5.4.4)

**Joints** Joints can be prismatic, revolute (rotary) or distributed. A distributed joint is an assembly between two rigid members which enables one to translate and/or rotate in relation to the other about an axis linked to the translation.
(ISO TR8373: 4.5)

**Learning control** A control scheme whereby experience (obtained during previous cycles) is automatically used to change control parameters and/or algorithms.
(ISO TR8373: 6.3.5)

**Manipulating industrial robot (MIR)** An automatically controlled, reprogrammable, multi-purpose, manipulative machine with several

degrees of freedom, which may be either fixed in place or mobile for use in industrial automation applications.
(ISO TR8373: 3.3)

**Manual data-input programming** Program generation and entry placed directly into the robot control system by means of switches, plug boards or keyboards.
(ISO TR8373: 6.2.2)

**Mechanical interface coordinate system** A coordinate system referenced to a robot's mechanical interface (the tool attachment point).
(ISO TR8373: 5.4.3)

**Sensory control** A control scheme where robot motion or force is adjusted in accordance with outputs of external sensors.
(ISO TR8373: 6.3.3)

**Teach programming** Programming performed by manually leading the robot end-effector, by manually leading a mechanical simulating device, or by using a teach pendant to move the robot through the desired motions.
(ISO TR8373: 6.2.3)

**World coordinate system** A coordinate system referenced to the Earth or a fixed position on the shop floor.
(ISO TR8372: 5.4.1)

**Wrist (secondary axis)** An interconnected set of links and powered joints between arm and end-effector which supports, positions and orientates the end-effector.
(ISO TR8373: 3.4.3)

## APPENDIX 2.B JAPANESE INDUSTRIAL STANDARDS (JIS)

JIRA adaptations of the ISO standard are:

1. glossary of terms for industrial robots (JIS B 0134–1986);
2. symbols for industrial robots (JIS B 0138–1980);
3. standard form for indicating characteristics and functions of industrial robots (JIS B 8431–1988);
4. measuring methods for characteristics and functions of industrial robots (JIS B 8432–1983);
5. general code for safety of industrial robots (JIS B 8433–1986);
6. identification symbols and colours for the operator controls of industrial robots (JIS B 8434–1984);
7. general code of design for modularization of industrial robots (JIS B 8535–1986);
8. circular flanged mechanical interfaces of industrial robots (JIS B 8436–1989).

Japan Industrial Robot Association Standards (JIRAS) cover:

1. standardization of electric signals of industrial robots (JIRAS 1006–1986);
2. standardization of piping between machines and industrial robots (JIRAS 1008–1986);
3. standardization of wiring between machine and industrial robot controllers (JIRAS 1009–1986);
4. standard interfaces between machines and industrial robot controllers (JIRAS 1010–1986);
5. holders of industrial robots.

Current standardization comprises:

1. standardization of robot language (ICR/high-level language);
2. robot message system.

# Introducing manufacturing systems and CIM

# 3

*Okitoshi Tsunoda*

Manufacturing systems and computer-integrated manufacturing (CIM) are closely associated with large, advanced production industries (automobiles, computers, semi-conductors, high-precision machinery, ships, aircraft and buildings). Recently the business environment in Japan has dramatically changed owing to increased competition and customer selectivity. To cope with this demanding business environment, it is essential that enterprises increase product reliability and shorten development and delivery times.

CIM technology is a key to success in meeting customer requirements. The use of CIM in production processes and business, in recent years, has expanded to include small industries and service industries. This chapter provides an overview of CIM and the methods for introducing it to users.

## 3.1 COMPETITIVENESS AND PRODUCTIVITY

To cope with today's demanding business environment, the world's leading industries have to be aware of 'what the customer wants'. Circumstances surrounding production activities and requirements in manufacturing are going through a period of radical change. Causes include the stronger yen, wider marketplaces, keener competition, diversification of needs, and increasingly rapid technological innovation. Management is faced with a wide variety of tasks and problems. To tackle them, and to obtain maximum customer satisfaction in terms of quality, cost and delivery time, management must effectively func-

*Manufacturing Automation Systems and CIM Factories*. Edited by K. Asai, S. Takashima and P.R. Edwards. Published in 1994 by Chapman & Hall, London. ISBN 0 412 48230 4

tion with the three key factors of engineering, manufacturing and marketing working together in close harmony.

### 3.1.1 COMPETITIVENESS

Looking at business trends, there are many elements needed to enable enterprises to respond smoothly to changes in the market, to increase the market, and to offer competitive products and service at the right time. Now, in Japan, leading manufacturing industries are promoting the production of higher quality manufactured products to meet customer requirements and play an important role in the world economy. The base of this rapid growth is the development of the manufacturing industries. Many industries have succeeded in supplying higher quality products, at low cost and with short lead time, through the use of manufacturing technologies. Furthermore, the study of competitive production is being carried out to globalize production by Japanese enterprises.

Japanese management, therefore, holds CIM as the key to the success of their competitive power-differentiation strategy. To be competitive, both product technology and production technology should be developed simultaneously and introduced to production.

### 3.1.2 PRODUCTIVITY

Productivity is defined as output per worker-hour, and is usually expressed in units manufactured per worker-hour invested. Using this measure of labour productivity, it seems that many different approaches can improve the productivity of workers having different skills and motivation. Looking at the productivity of production, it may be represented as influenced by the key factors shown in Figure 3.1. Examples are:

(a) Manufacturing industries – form conversion: materials and information are converted to valuable assets.
(b) Distribution industries – place conversion: value is actualized by the process of changing the asset's place.
(c) Information service industries – time conversion: value is added by the process of consuming time.

In production activities, efficiency of production is defined as productivity, and productivity is required to satisfy customer needs. To realize high productivity, the integration of 'hard' and 'soft' technology, called manufacturing technology, is very important. Hard technology means 'the production process flow of converting material to product'. Soft technology means 'production management to control production

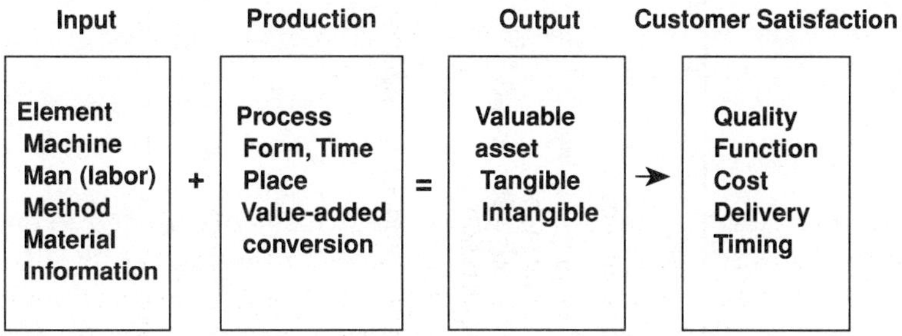


Figure 3.1 Key influences on productivity.

information'. The manufacturing technologies mentioned above have been developed to improve productivity and practicality, at reasonable cost. Generally speaking, efficiently improved production technologies should be incorporated into production with management priority as a combination of 3M plus I (man/material/method and information). In the future, attention must be paid to improving the speed of production; information, therefore, becomes more important for improving productivity.

### 3.1.3 HISTORY

Before describing a method for introducing manufacturing systems, it is worth looking at the history of the production process in order to understand the innovation trends for industrial manufacturing technology (Table 3.1).

Japan is currently leading the world in the possession and manufacture of industrial robots (IR); approximately 200 000 robots are in service, and production reached 41 000 per year in 1984. Moreover, the application of industrial robots is under consideration in all industrial fields regardless of size. It was recently reported that half of the world's total robot population is in Japan. Previously, the application of IR was limited to mass production, difficult working environments and heavy industries, especially automobiles and consumer electronic goods. Particularly noteworthy for mass production is the fact that not only IR but also large-scale automation was required, and many other technologies were needed to keep product cost low. Thereafter, the introduction and proliferation of IR and automation have been rapidly promoted in all industrial fields.

The original reason for the development and utilization of IR was to replace humans in dangerous or monotonous work. The installation of IR over the past 15 years, however, has changed from a stand-alone

**Table 3.1** History of production technology progress

| | 1960 | 1970 | 1980 | 1990 |
|---|---|---|---|---|
| Business | High growth | | Steady growth | Reasonable growth |
| Purpose | Direct productivity<br>High-volume production<br>Low cost<br>Labor saving | | Indirect productivity<br>Diversification<br>Higher quality<br>Soft strength | Business productivity<br>Personalization<br>Higher business speed<br>Globalization |
| Technology | Industrial robot<br>Partial FMS<br>Hard automated equipments | | FMS<br>Automated process<br>Factory automation<br>CAD/CAM<br>MRP | CIM<br>Integrated production systems<br>Network<br>AI |
| Computer | Host computer | | Microcomputer<br>Personal computer | Workstation<br>Handy computer |
| Control form | Stand-alone | | Island | Integration |

to a systematized mode, and improvements in IR have caused the expansion of IR applications. The step now being taken is to apply and to install IR on a large scale in fully automated production lines and FMS. Hardware and software technologies have been rapidly innovated in computers and systems, and their cost-effectiveness has been dramatically improved. Semiconductor technology, in particular, influences the entire computer industry (microcomputers, personal computers and workstations). Communications technology has also been dynamically improved. Network application is currently being expanded both within individual factories as well as in worldwide computer applications.

Recently, many consumers have started to demand personalized or diversified products. This requires great changes if companies are to stay competitive because not only manufacturing productivity, but also marketing and engineering productivities must be improved at the same time. In addition to productivity, management is focused on a 'manufacturing strategy' based on a new marketing concept for consumers. From the manufacturing point of view, the production concept should be changed from 'product out' to 'market in', and the productivity improvements needed at the company-wide level are shorter total lead time, higher quality and wider variety. CIM is now essential to achieve the production efficiency needed to meet these company-wide improvements.

### 3.1.4 OPERATION IMPROVEMENT

At the beginning of each year, top management directions cascade down to each line manager, who must then establish individual targets for improvements based on the findings of the previous year. These manufacturing targets are incorporated into all operational improvements including total quality control (TQC) and other methods. All managers define measurement and techniques which determine measurable targets for improvements in process efficiency.

For more competitive manufacturing, two key objectives must be met:

(a) elimination of all manufacturing process waste;
(b) establishment of a manufacturing architecture for the production.

These can be achieved with continuous flow manufacturing (CFM) activity, initially introduced in the assembly and test processes. With continuing effort and experience, CFM will be expanded into other areas. The basic operational improvement that must be performed is to optimize and simplify the operations between functions. During the process of improving operations, talented people will establish

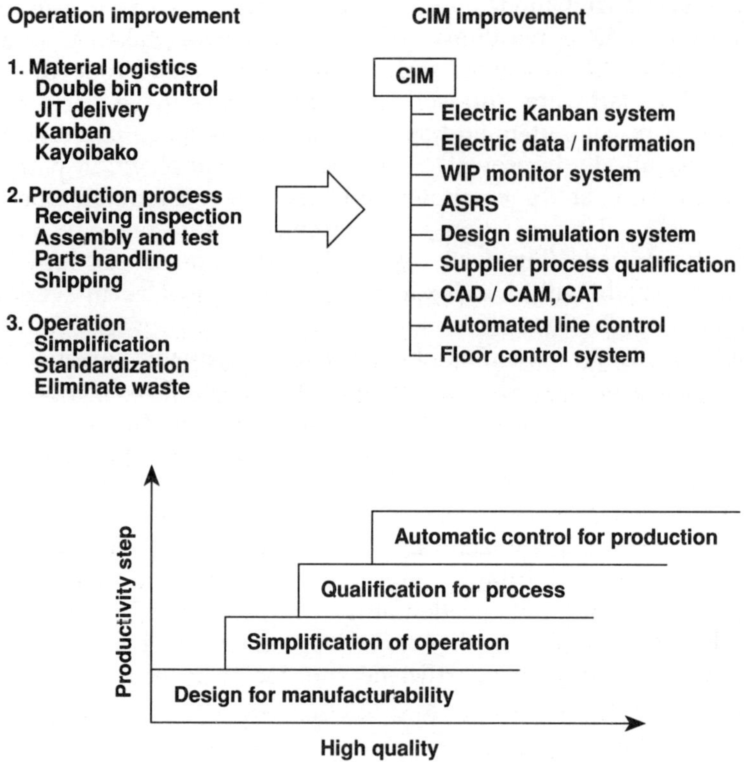


Figure 3.2 Introduction of CIM for operation improvement.

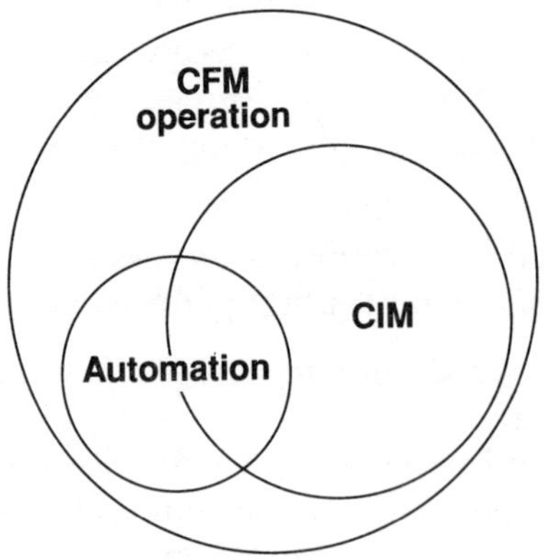


Figure 3.3 Integrated CIM/CFM automation.

innovative improvements using automation and other techniques. They must evaluate whether the production processes offer competitive advantages for meeting present and future goals. For improving production efficiency, CIM is introduced in an order determined by considering the priority of each step for new products, business trends, production process, and also technological innovation (Figure 3.2).

## 3.2 THE CIM CONCEPT

A basic concept for introducing CIM in manufacturing industries is the importance of achieving higher productivity from the working combination of CIM/CFM/automation (Figure 3.3). To increase business efficiency, a company constructs its CIM in organic combination with marketing, engineering and manufacturing (Figure 3.4).

### 3.2.1 CIM OBJECTIVES

It is now five years since CIM was introduced into Japan. For manufacturing in the United States to become competitive using CIM, it must become one of the nation's prime considerations, which is an example of the top–down approach. However, in Japan, attention must be directed toward the manufacturing process and its improvement by factory automation (the bottom–up approach).

In manufacturing industries, the most automated forms of production are computer-controlled equipment and manufacturing systems. A variety of names have been given to these systems, including FMS,

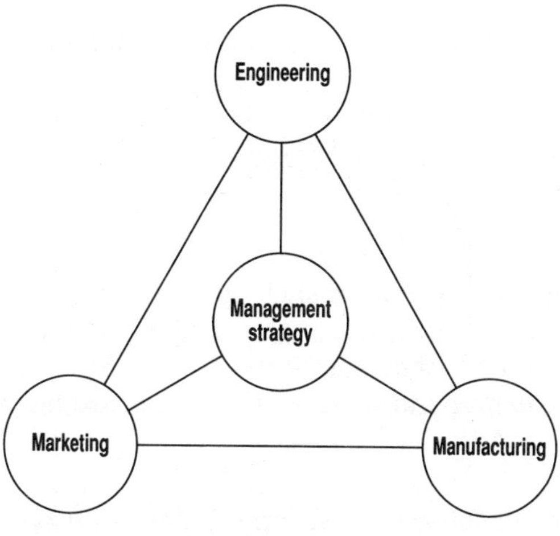


Figure 3.4 The concept of CIM.

| | | | |
|---|---|---|---|
| **Leverage** | **Product superiority** | **Manufacturing superiority** | **Business superiority** |
| **Production** | **Small volume** | **Medium volume** | **Large volume** |
| **Achieved by** | **Design** | **Flexibility** | **Automation** |
| **Operation** | **Engineering-driven** | **Engineering/ manufacturing driven** | **Manufacturing driven** |
| **Mechanization** | **Uniqueness** | **Manufacturability** | **Automationability** |
| **Type of result** | **Features** | **Customized** | **Low cost** |
| **Competitiveness** | **Functions** | **Technology** | **Cost** |

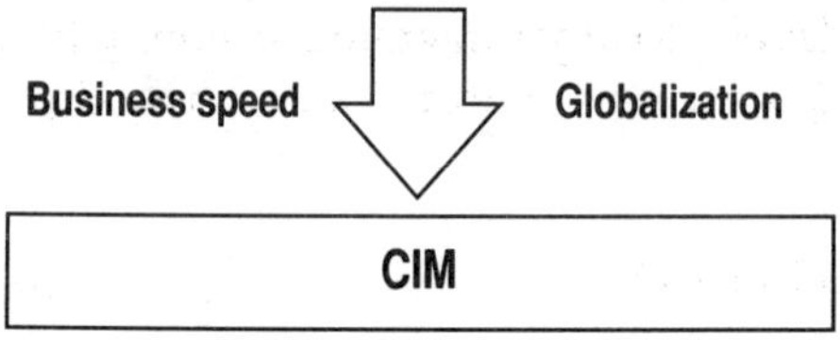


Figure 3.5 Competitiveness approach for CIM systems.

CAD/CAM, CAE and office automation. From now on, we shall categorize all these systems as CIM. For upgrading the production process, the introduction of CIM is categorized by each business type as regards production performance and product leverage (Figure 3.5).

### 3.2.2 CIM SYSTEMS

For enhancing production into a manufacturing system, CIM is introduced at each step according to priorities determined by consideration of economical, periodic and circumstantial changes, and also technological innovation. The following two points will need to be considered when constructing a CIM system to achieve higher productivity from a competitive point of view.

#### (a) System integration

To cope with business variables such as quality improvements, fluctuations of production volume, improvements of indirect work, and speeding up business, the most important factor is the CIM system architecture regulating information flows. Regarding higher productivity of indirect work, the following aspects are required to introduce the CIM system:

(a) Management of job-classified information – introduce hierarchical systems to executives, managers, engineers and workers at each level; vertically integrate from top to bottom.

(b) Integrated control of database – introduce a common database and integrate horizontally systems such as factory automation, OA and their staff.
(c) Enlargement of network – introduce standardization of information exchange and communication (network integration of island systems).

CIM can be established effectively by step-by-step implementation into the following project construction levels:

(a) Island level – individual systems or stand-alone systems for CAD and MRP.
(b) Interface level – individual systems with data exchange via network and database.
(c) Integration level – totally integrated system with common database, network, engineering applications and linked production lines.

### (b) CIM system hierarchy

In general, the functions accomplished by the computer control system can be divided into five layers. The following descriptions of the system hierarchy (see Figure 3.6) apply best to practical manufacturing systems.

(a) Factory management system: This information processing system supports administrative activities and decisions of factory management according to information and real records obtained from the factory control system (Layer 2), from comparison forecasts adjusted monthly according to real records, etc.
(b) Factory control system: A corporate interface is the group of systems coordinating the international information network for the global operation of branch offices, laboratories and factories. It comprises the technology information system, the order-received information system, and the production schedule information system. The factory administrative control system comprises purchase, accounting, personnel information systems, etc.
(c) Engineering systems and production control systems: Engineering systems support engineering activities such as CAD/CAM, CAE and CAQC, as well as process design and simulation analysis. The production control system has many functions: material procurement scheduling, materials order control, and inventory control centring on MRP.
(d) Production-floor control system: A shop-floor control system takes care of process control and materials handling. It also controls production lines, and monitors progress, production quantities and quality data.
(e) Production-line control system: This system controls production

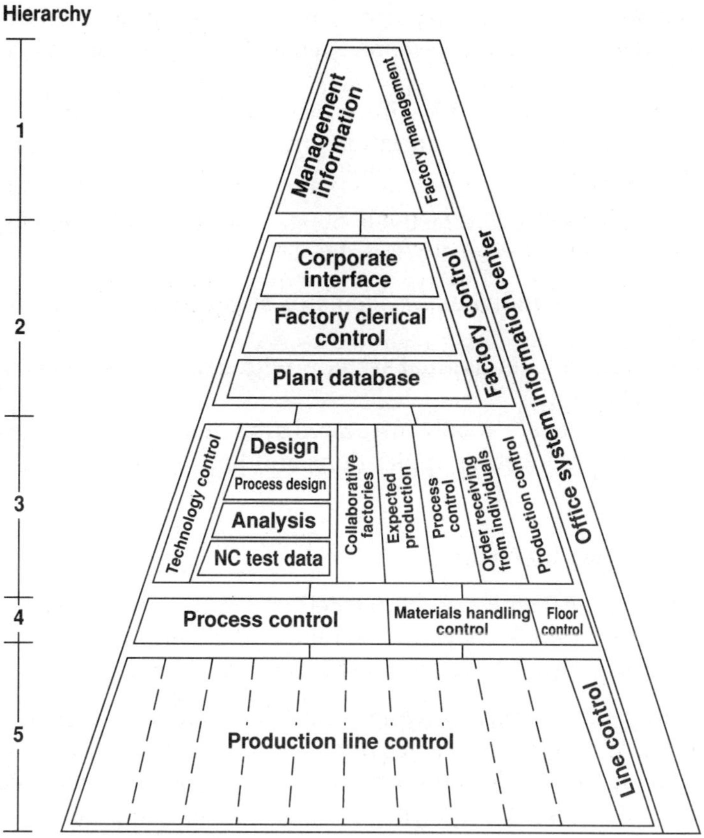


Figure 3.6 CIM hierarchy.

equipment such as transport devices, automated warehouses, conveyors, machine tools, testing equipment, inspection machines, and assembly robots in the production line. In many cases, the line control system connects via networks with the production-floor control system and transmits equipment information from its controllers. These systems are called CAI, CAT and CAA.

In addition to this system hierarchy, it is essential to focus on productivity in terms of quality, cost and delivery.

### (c) CIM technology (systematization)

In order to produce with high efficiency, material and information from computer-controlled equipment can be combined without delay by

CIM technologies such as JIT operation, monitored production and defect analysis. The most important aspect here is the CIM systematization in handling flows of materials. For higher productivity, two forms of the CIM technologies are very important:

(a) **Technical information management** – rapidly processes materials and information by information processing functions and integration of technical information such as:
   (i) monitoring and reporting on progress conditions;
   (ii) analysing and reporting on quality conditions;
   (iii) control of individual level progress;
   (iv) quickly changing schedules in response to activity;
   (v) filing of historical changes.
(b) **Automatic accumulation** – automatically collects numerous data and analyses information by system functions such as:
   (i) modular systems which can be reproducible;
   (ii) recognition of equipment operating conditions;
   (iii) automatic connection of each process;
   (iv) feedback control of information from later to earlier processes;
   (v) job-classified computer control of various equipment.

The objective is for each factory activity, including fundamental functions such as production schedule management and production, to be executed as a continuous flow without human intervention and with recognition of production activities when necessary.

The current trend is for the form of production not to be a fixed format, and more flexibility will be required in production. To achieve CIM suitable for each factory, automatic connection and management of job-classified information are essential.

### 3.2.3 BASIC IMPLEMENTATION STEP

Many users who want to install manufacturing systems seem to be under the impression that they can obtain immediate productivity improvement merely by purchasing cheap equipment after a hasty selection. In cases where implementation is easy, the production system might be installed without any problem, but it seems unlikely that maximum customer satisfaction can be obtained generally without considering the unique requirements of each case. Therefore, manufacturing project plans including IE jobs, production control jobs and accessories work-flow, are indispensable for planning line works or unmanned FMS by robots and automated equipment. Future changes should also be allowed for as far as possible.

In general, there are three basic steps in implementing manufacturing systems and/or production facilities:

(a) **Preparation and planning**. During the planning of a manufacturing system, it is essential that CFM (continuous flow manufacturing) be studied not only by engineers but also by manufacturing technicians. Established CFM operation provides the basis for automation and CIM.
(b) **Development and implementation**. For the development of automated equipment and systems, the most important thing is maintainability and reliability.
(c) **Improvement and full utilization**. To achieve high production objectives, all employees need to meet the challenge of productivity improvement with teamwork conducted by manufacturing operators and technicians, similar to a quality circle.

## 3.3 IMPLEMENTATION

An implementation approach giving CIM a key position in the success of the competitive power-differentiation strategy is shown below:

(a) Construct the CIM system using IBM's concept.
(b) Establish a CIM promotion division and promote CIM, CFM and automation as a combined improvement strategy.
(c) Integrate the top–down approach applied by the US and European industries with the bottom–up approach applied by Japanese industries.
(d) Develop and review the comprehensive CIM plan based on the enterprise business strategy at every plan cycle.
(e) Apply the ladder approach as step-by-step implementation, and define the implementation priority depending on the technological readiness.
(f) Improve the system hardware and software continuously to increase competitiveness.
(g) Promote network integration and a common database.
(h) Educate not only the system engineers but also the users and operators.
(i) Adopt security checks and audits.
(j) Establish the goal of completing a high-level CIM factory in the future.

The design and development currently needed to introduce CIM systems are more complex than in the initial development period. It normally takes a long time to establish CIM and, sometimes, the business environment will have changed by the time CIM is released. As mentioned above, it is therefore essential to implement the CIM systems in a strategic way, with careful review of each step.

### 3.3.1 THE CIM CONCEPT IN IBM

What is CIM? IBM defines CIM as a strategic information system for improving total productivity by the efficient integration of engineering, manufacturing and marketing. The benefits are not only functional efficiency but also total enterprise efficiency, and IBM offers a strategic information system to support the objectives shown below:

(a) management efficiency by the integration of internal and external functions;
(b) competitiveness reinforcement by speeding up business processes;
(c) internationalization by global operation.

There are two key aspects of the business environment of production. The first is demand fluctuation. Because it is difficult to project demand and supply according to market trends, the production plan and schedule must be updated on a short-term basis. Various production features are altered to meet changing customer requirements such as new specifications, different delivery dates, and order cancellations. The results of these changes may impact on many business aspects such as remaining inventory, substantial work-in-process, parts shortage, schedule delays and shipping stoppages. To manage these various changes, increasing control of the operation will be required. The second aspect is technological innovation. It is essential that new technologies penetrate into products in order to improve quality and reduce cost. Sometimes, engineering changes will even be applied to improve performance in existing products. Technological innovation also has a big impact on demand and supply; it may therefore cause considerable waste in the enterprise, and there may be no profit when the business processes are too complex and slow.

### 3.3.2 CIM PROMOTION

To progress the implementation of CIM, a project office or promotion division should be organized with capable managers and members, such as engineers and planners with good manufacturing experience. This project office is an independent function of the DP system, and always works as a combination of CIM, CFM and automation to improve the production process with Japanese methods such as JIT and flow manufacturing (Figure 3.7).

As examples for CIM promotion, targets to achieve competitiveness driven by top management are defined by TQC and other methods. The current year's target is cascaded down at the beginning of the year and a long-range strategic plan is also established by the project office. To meet the business target, the project office has to drive the implementation with user functions.

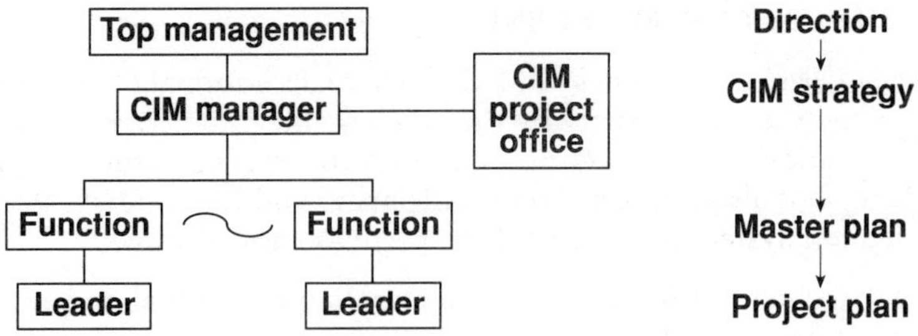


Figure 3.7 CIM organization.

### 3.3.3 TOP–DOWN AND BOTTOM–UP APPROACHES

Generally speaking, the world-leading industries in the USA and Europe are competing by means of a top–down approach, but Japanese industries are growing with a bottom–up approach.

| *Top–down* | *Bottom–up* |
|---|---|
| Management-driven | Manufacturing-driven |
| Organic method | Functional method |
| Business first | Quality first |
| Short-term merit | Long-term merit |
| Innovation | Improvement |
| Job hierarchy | Team work |

To realize a suitable production system and to support timely business changes, hardware and software are also rapidly enhanced. The most important aspect, however, is the involvement with people through motivational tools such as quality circles and Kaizen suggestion programmes. In many enterprises, a number of suggestions about quality circles are also incorporated as 'group' suggestions in addition to individual suggestions. Prizes are awarded in many motivational ways. Through such activities, the production system will be gradually improved by a bottom–up approach.

The top–down approach offers advantages for formulating a business strategy and for setting the target of the business with management priority. From the point of view of technical skill and system design, the bottom–up approach offers advantages for implementation with a feasible schedule. Therefore, Japanese attention is sometimes paid to improvement activities rather than the innovative activities. USA and European attentions are directed towards outstanding concepts and a large system plan, but the development lead-time based on that concept is sometimes too long. A combination of the top–down and the bottom–up approaches is essential to increase the productivity in each area.

### 3.3.4 CIM COMPREHENSIVE PLAN

The first question is: what is the target of CIM for the business? In the top–down approach to supporting the business strategy through the information process, a key question is 'How long?'. A long-range plan for CIM is the same as the long-range strategic plan of the enterprise; it is probably of a few years' (five or six) duration. A proposed approach is as follows:

(a) **Give a direction from business strategy**. The improvement plan must be changed to a strategic plan. Functional organization must be changed to strategic organization.
(b) **Promote integration of each function**. Partial improvement must be changed to total improvement. Individual decisions must be changed to a totally balanced decision.
(c) **Take quick actions and make quick decisions**. Individual actions must be rapidly changed to wide-ranging actions in order to affect total performance.

The CIM comprehensive plan should be developed with a broad view based on the long-range business strategy and plans shown in Figure 3.8.

Figure 3.8 Top–down approach.

| *Plan* | *Review cycle* |
|---|---|
| A long-range strategic plan | Once a year |
| A mid/long-range CIM plan | Twice a year |
| Implementation plan | Once a year |
| Development and implementation progress | Once a week as required |

Normally, the CIM project plans are developed by a well-organized team with divisions such as promotion, planning, production control and manufacturing. Management of the enterprise must review the plans and the progress and, if required, must take quick action.

### 3.3.5 STEP-BY-STEP INFORMATION

The next topic is how to implement, or how to realize CIM. A key aspect is step-by-step implementation, as the bottom–up approach is based on a long-range plan for CIM.

To establish the CIM implementation plan, the following actions should be considered:

(a) study technical feasibility or availability;
(b) estimate investment of capital and resources;
(c) define challenges of tasks and problems;
(d) promote implementation of beneficial areas;
(e) improve the system for full use;
(f) enhance the system for business expansion.

After discussion of the above aspects with the relevant members, the implementation steps will be decided by a CIM manager. In Japan, the three-step approach (hop, step and jump) is very popular for implementation. 'Hop' means a trial implementation to assess the process; 'step' means regular implementation to pursue production; and 'jump' means some innovative implementation to make the business grow. Associated with the business growth, the system enhancement will be promoted more strongly to improve all operations. An example of CIM enhancement is shown in Figure 3.9.

### 3.3.6 CONTINUOUS IMPROVEMENT

It is well understood that better quality assurance is the key to quality control. Manufacturing must be able to ensure that products will satisfy customer needs in terms of performance, quality and reliability while still in the manufacturing process. If system reliability is to be improved, the productivity improvement of a factory cannot be achieved by com-

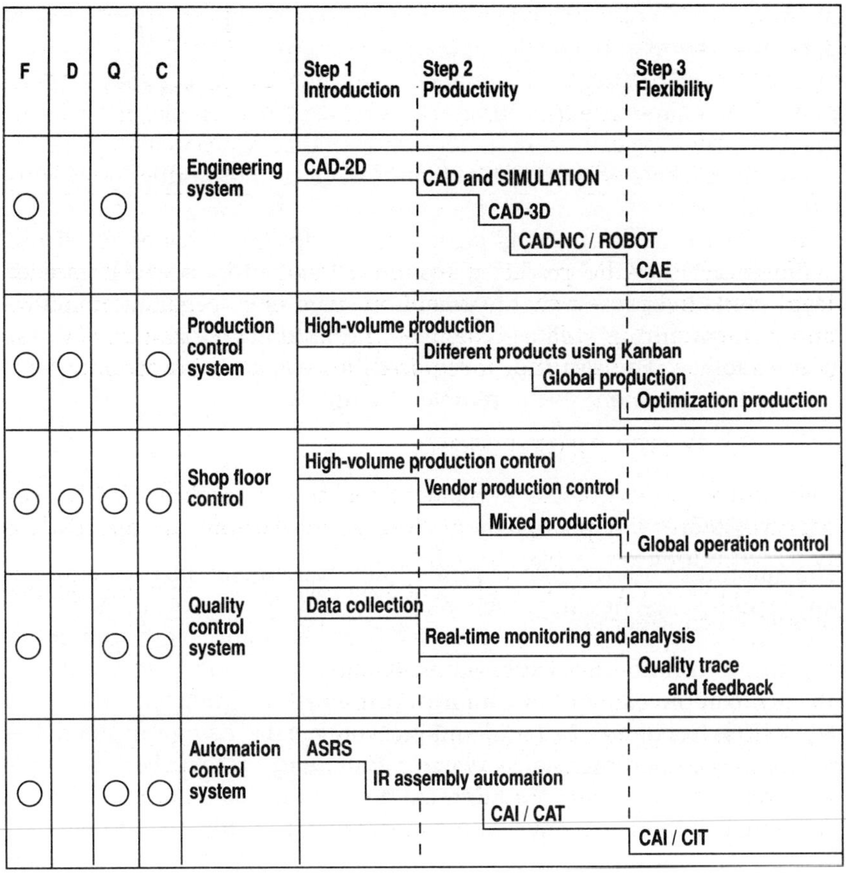


Figure 3.9 CIM enhancement.

mercial equipment. Rather, continuous hardware and software improvements are essential. Through experience with such continuous improvements, members are provided with a good opportunity to learn about leadership.

Manufacturing is involved in the development of products by the investigation of manufacturability and assemblability. Without this involvement, it is difficult to be competitive and to reduce problems in the factory. During the design phase, intensive problem-solving efforts are made prior to releasing the CIM to a production line. One good example is a team effort called the quality circle. This voluntary team works on management, data analysis, technique improvement and communication, as well as on the presentation of activity achievements. The team initiates a way to solve a problem or to propose a problem fix by using QC tools. In general, 'first aid' on equipment problems is performed by manufacturing operators.

In addition to their maintenance jobs, maintenance people spend part of their time in supporting engineering. For example, in order to prevent accidents caused by faulty maintenance or service after installation, a fail-safe system of precaution or a monitoring system is essential.

Manufacturing technicians participate in design activities to ensure maintainability of the products, equipment and processes. Along with their efforts to ensure a highly reliable system, engineers and manufacturing operators, as well as managers, are challenged to improve system reliability. The result of these activities will be a practical system; plus an improvement in the morale of employees.

### 3.3.7 NETWORK INTEGRATION

The amount of information required to do production jobs in manufacturing industries has increased recently, and an improvement in the productivity of information processing is becoming more and more necessary. Information processing also has to be 100% accurate and rapidly completed, and all information must be transferred as and when it is required. Network integration means communication between people and people, between people and system, between system and system, between equipment and system, and so on.

Normally, the communication functions and systems are integrated by networks into, effectively, a single system that can communicate between any location, any system and any application on the networks (Figure 3.10). The following two networks are essential to support internationally diverse production and collaborative production:

(a) **International network**. Products developed in Japanese laboratories are produced not only in domestic factories, but also in overseas factories; and products developed in overseas laboratories are also produced in Japanese factories. This divided production system, based on the concept of producing superior products in areas most suitable for the production, is one of the most noticeable in the world.

(b) **Local area network**. To produce a superior product, many direct and indirect functions must work in close operation. An office automation LAN provides timely information to indirect people, and an industry LAN keeps production and management informed on the status of the production line. Communication via local area network can result in quick action.

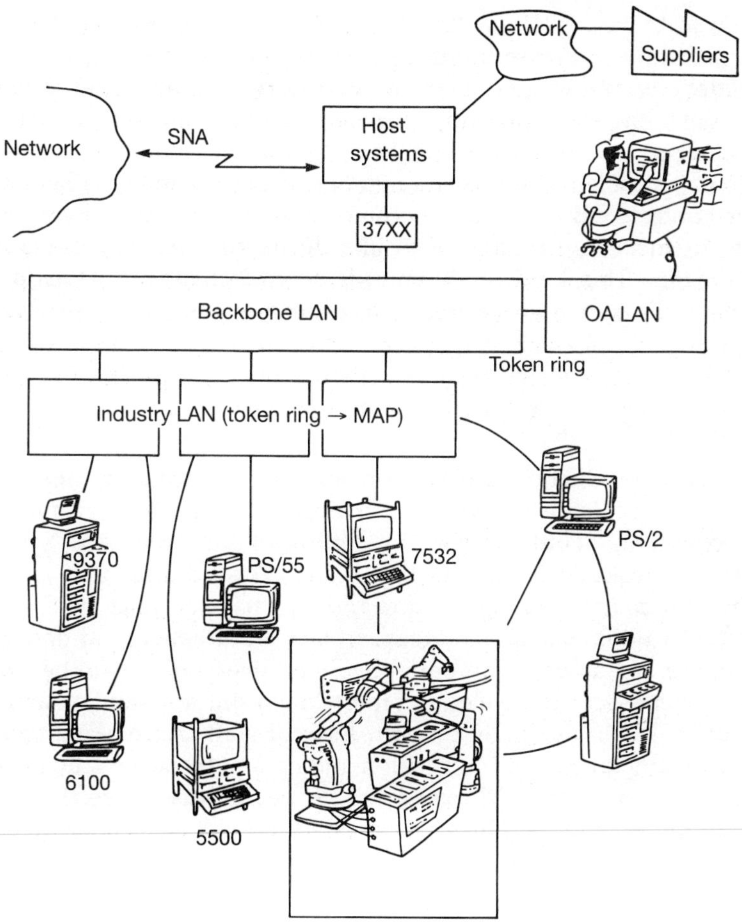


**Figure 3.10** An integrated network.

### 3.3.8 EDUCATION

True CIM cannot be achieved overnight; it will take years of continuing efforts in developing, manufacturing and engineering. To achieve practical CIM, it is important that the people at the top of the system learn the value and strategy of CIM. The generic and technical skills of employees must be enhanced by career-planned education programmes. Education and training programmes will enhance broad skills including delegation and motivation. It is essential to encourage employees to do intensive job assignments like operator maintenance, and to become involved in development during manufacturing. Recently, manufacturing employment in industrially advanced countries has fallen year by year because the work is difficult, dangerous or dirty. To cope with this severe environment, employees can no longer depend on somebody else to give them the necessary education and training. They must, therefore, develop themselves, acquiring new knowledge and studying on their own initiative. As the operators begin solving day-to-day problems and as the manufacturing discipline is kept well under control, engineers can become more flexible and divert their efforts to more constructive jobs like developing new products and technologies. This kind of situation is a good chance to develop the various skills and manufacturing technologies needed to survive in business.

### 3.3.9 SYSTEM SECURITY

To increase production information and data processing simultaneously, it is necessary to create an environment in which everyone is more conscious about the security control of the processing system. Generally speaking, the production system has been tailored to meet user requirements, and information and data have become increasingly important to this business strategy. In fact, information and data processing are key assets in an enterprise, and everyone has to be aware of the system security needed to protect them. Information is classified and controlled individually by each level of staff. Authorized system users can only access the information and data of the system by using a personalized password. These days each person may have their own office terminal, and data files are accessed individually under the control of the system security.

### 3.3.10 ESTABLISHMENT OF THE CIM GOAL

To be competitive, an enterprise must always be concerned about quality, cost and delivery improvements. Enterprises that promote company-wide programmes involving top management as well as everyone else establish systems that provide high-quality products to customers worldwide. As the 21st century approaches, each enterprise must establish the goal of having the best CIM.

## 3.4 FACTORY AUTOMATION AND CIM SYSTEMS

CIM is a practical manufacturing system for manufacturing various products using mathematical methods. Manufacturing industries now have several issues to be solved as the business environment changes:

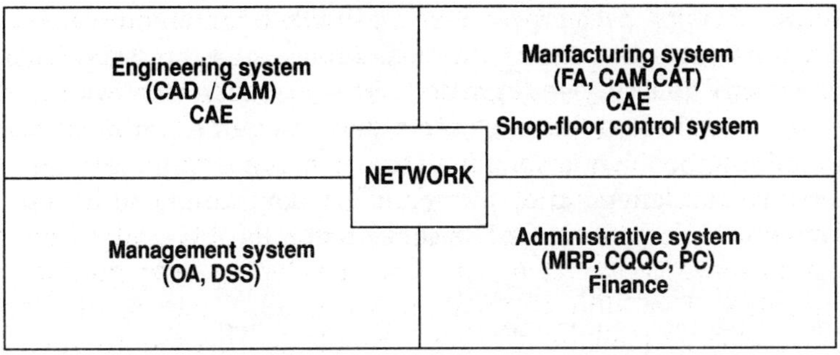


**Figure 3.11** CIM elements.

(a) decreased manufacturing resources;
(b) increased complexity and variety of production processes;
(c) improved reliability of equipment and processes;
(d) increasingly divided production;
(e) increased importance of information for production.

These issues can be solved by a CIM system flexible enough to establish a variety of system designs (Figure 3.11). The various elements of CIM include CAD/CAM, CAE, factory automation, CAT, OA, DSS, MRP and CAQC, all controlled by computers and/or by computer applications. Improving the manufacturing process or solving manufacturing problems are fully dependent on the integration of many such systems, as described in the following sections.

### 3.4.1 CAD/CAM SYSTEMS

In the past decade, CAD/CAM technology has provided the impetus for the growth of Japanese productivity. It has developed in response to industry needs for sophisticated interactive graphic terminals, computer-controlled equipment, intelligent robots, automated inspection and other innovations to produce higher-quality products with higher productivity.

This CAD/CAM technology integrates well into the production environment in Japan because, in the past, design and production were treated, without distinction, as being part of the same function, in Japanese factories. Most development engineers have experience of manufacturing, and this experience will ensure the awareness of quality and attention to the underlying technology needed for production using computers and tools. The technology of CAD/CAM should be advanced from the standpoint of the historical business environment.

CAD is defined as the use of computers for designing, with graphics for drawing, with data processing, and with information checking. It has been useful and meaningful for designers, and the result has been better design quality. Traditionally, design quality was evaluated by a professional through his experience and through experiments. However, this method has a long lead time, requires a large number of professional designers, and can result in poor manufacturability. The reasons for implementing CAD systems are:

(a) **Increased design productivity**. Designers must produce designs to meet customer requirements within a specified period.
(b) **Improved design quality**. It is essential to reduce design errors though engineering analysis and experience in order to make products with built-in quality.
(c) **Evaluation of design for manufacturability**. In the process of designing products, the database required for trial manufacture is created. The design can thus be evaluated quickly by computer simulation or by sample products.

When the CAD system is installed, the users must examine the various hardware and software components that make up a useful and beneficial system. The purpose of the CAD system should also be clarified with a long-range strategy for future products. The functional elements required for CAD can be grouped into four categories:

(a) **Operationability**. Interactive graphics terminals and other components must provide user satisfaction in terms of usability, visibility and information exchange.
(b) **Functionality**. Modelling, 3-D drafting, analysing and designing should be sufficiently and flexibly provided.
(c) **Expandability**. System enhancement must be flexible in order to provide hardware and software components as needs arise.
(d) **Connectability**. The system must be able to transfer engineering data via networks from any system to any other system.

It is important that users pay careful attention to these categories because of the long-term effects of new production in terms of quality, cost and delivery. Another important aspect is the ability of users to design and plan for development through education, training and experiment.

CAM can theoretically be defined as the use of a computer to plan, manage and control all factory operations. But in Japan, CAM is focused on the production process, so the term 'FA' (factory automation) is very popular. Applications of CAM fall into broad categories: many types of products, small- and large-scale production, short life-cycle, technologically innovative products, etc. In addition to supporting production, the CAM system also indirectly supports manufacturing with computers providing administrative job improvement. The CAM system is not only linked directly to the production process but also to the office operations so that the manufacturing operation can be managed effectively.

Production facilities used to depend on the scale of production volume; the production objective was simply to generate products for sale. Recently, however, customer-driven manufacturing has rapidly grown to produce products demanded by customers. This production also aims for products which have a good sale. To achieve customer-driven manufacturing, CAM was promoted and implemented by all industries in their step-by-step approach towards CIM.

| CAM step | Manufacturing focus |
|---|---|
| (1) Partially automated tools | Operation improvement in parts manufacturing |
| (2) Automated equipment and facilities | Cost-saving in the assembly process |
| (3) Fully automated process | Competitiveness for large production |
| (4) Total integrated process (CIM) | Flexible and new production enhanced by the CAD/CAM system |

With the evolution of machines, CAD/CAM, and other computer-controlled systems, manufacturers are moving toward unmanned factories in the future.

In Japan today, the manufacturing technologies most focused on production are the system integration technologies:

(a) Hardware – intelligent equipment, robots, and automated facilities;
(b) Software – intelligent CAE, man–machine interfaces, and a common database;
(c) Systems – flexibility, compact size, low cost, and high reliability;
(d) Sensors – precise, multi-function, and sensing methods;
(e) Networks – high speed, low noise, and low cost.

### 3.4.2 NETWORK AND FACTORY AUTOMATION COMPUTERS

Since the early 1980s, the installation of assembly robots in Japanese automation lines has become increasingly popular. The assembly process is an in-line stand-alone system consisting of a main conveyor, robots and a controller. This automation, initially named factory automation, is included for the mass production of consumer products such as cassette tape recorders, electric fans, VTR mechanisms, PCBs and watches. The in-line system is designed as a sequence of numerous simple operations, including parts quality improvement and JIT operation. There is only a physical interface between hard automation and the robots, and the automated process is only loosely coupled with information processing.

Recently, other requirements have been batch raised by mixed production, which features large variety and small size; moreover a rapid and flexible response is needed to meet changes in production schedules. This is made possible by the integration of automation technology and computer systems designed to use information networks. Furthermore, it is desirable to improve both the efficiency of each process and the overall efficiency of the production processes as a whole. Accordingly, there has been an increasing role for CIM in the automated process as the needs for both the technical and production information have increased.

For the automation of high-quality assembly of precision parts, timely monitoring and acquisition of current progress status are required. Future manufacturing is likely to become even more precise, complex and compact. To meet future needs for maximizing overall assembly efficiency, the trend must be towards computer-integrated assembly in order to deal with intelligent robots and networks.

Today, the various small computers often used in equipment controllers, personal workstations and microcomputers, as well as programmable logic controllers or sequences, are popularly called factory automation computers in Japan. They have two functional characteristics: process control computing and data processing at the shop-floor control position. The main function of the factory automation computer is data management by communication with a host system, but another important function is to respond quickly to production data for the PLCs or the controllers housed in robots and other automated equipment. Making a new line, or modifying the line in a short time, requires a flexible computer system for changing, upgrading, and adding functions. For maximum productivity, a production line must operate non-stop; so the system reliability must be improved by using a computer system and a monitoring system. To be useful, the factory automation computer has user interfaces such as digital-to-analogue converters and standard bus lines, so that users can connect any devices, measurement tools, and/or other equipment.

In large and complex systems, a computer hierarchy must be defined to control the production activities in a factory. The general hierarchy levels of the computer system, which were shown earlier in Figure 3.6, are as follows. At the fifth level of computer control, computers are connected directly to the production process. They are PLCs or microcomputers dedicated to process control and they communicate with the factory automation systems through a LAN in the fourth level minicomputer used for shop-floor control and progress monitoring. The third level of computer control consists mainly of production management systems for MRP, planning and scheduling, and the technology control systems named CAD/CAE. The second-level computer is a large central computer which has numerous interfaces to communicate with other systems, and also processes common applications. The first-level computer system is a management computer supporting administrative activities and the decisions of factory management.

In order to ensure the effectiveness of factory automation, managers and owners define performance measurements such as yield monitoring, work-in-process, production scheduling and equipment availability. In general, it is standard practice to set up networks or interfaces to collect automatically the following information and data:

(a) very high speed and control data transmission – common bus line;
(b) large data volume and high-speed transmission – local area network;
(c) low-speed and long-distance transmission – special telephone line;
(d) international transmission – international telephone network.

In general, the LAN is installed as a computer–computer communication interface and industrial robots and other machine controllers are equipped with special or standard interfaces with minicomputers or microcomputers. As an automated assembly line, each robot has a common interface and is connected with a robot controller for line control. The interface of a robot network is linked to a local system or network for the shop-floor control system, and to a host system for total production control and factory management.

In addition to the above interfaces, a corporate interface is one element of the system that allows the total information network to coordinate the global operations of laboratories, branch offices and head office. It is important to speed up the clerical processing of administrative information and data. In Japan, this supporting system, named office automation (OA), is brought into the CIM system via an OA LAN. It improves office work productivity including that of the management office. The system utilization must be as close to 100% as possible, especially for OA terminals that have multiple functions.

Any CIM system design projects must be organized to be effective. During the design and development stage, revision tends to effect a wider range of areas than expected. It is, therefore, essential to establish a standardized interface in a modular system, and all systems in a factory must be integrated vertically and horizontally by the common interface or network.

### 3.4.3 INDUSTRIAL ROBOT LINES

An industrial robot is a programmable general-purpose machine with certain human-like characteristics. It can be programmed to perform a sequence of motions to control a variety of operations. In Japan, robots are popularly and economically used in assembly processes. Recently, there have been many innovations in the conception and design of production lines and systems. Until now, when designing a system for the application of industrial robots to a new production system, feasibility and adaptability have been studied either by considering the functions and cost of various industrial robots already in use, or else by considering a prototype from the manufacturers. However, this selection approach has been reconsidered owing to the application of robots to more complicated products. Additionally, not only are simple robots now available but also one to six control-axis robots. Sensor-equipped robots are becoming more popular. For precise control of position, a machine-vision system can also be attached to a robot.

Basically, industrial robot selection considers the following points:

(a) configurations – polar coordinate, cylindrical coordinate, jointed arm, cartesian coordinate;
(b) functions – speed control, weight carrying, positioning control, work area;
(c) features – interface, sensors, vision control, grippers, programming, teaching;
(d) safety – hardware and software protection, monitoring.

For example, in handling work for top cover mounting and if minimizing the work locus, the control axis is planned by combining a conveyor, pallet fixture, automatic parts feeder, etc. This job can be accomplished with only three axes. In general, much handling or assembly can be dealt with by industrial robots having fewer control axes.

The development of this process with respect to the in-line robot assembly is:

(a) an equally divided sequential process;
(b) a simplified operation by assemblability evaluation;
(c) a balanced cycle time of each process;
(d) a common pallet and conveyor as a combination.

The application system for assembly robots requires a readjustment in thinking relative to the new system design method, in which the necessary locus is analysed with the goal of minimizing the number of control axes in order to minimize cost. For example, one proposed approach for designing and implementing an industrial robot assembly line is as follows:

(a) define in-line type or cell type depending on production;
(b) design divided processes based on cycle time;
(c) design layout of all processes with materials handling;
(d) evaluate process design using computer simulation;
(e) select the industrial robots and facilities;
(f) develop the application system with users;
(g) implement the robot assembly line within a short lead time;
(h) improve the robots and the system to maximize effectiveness;
(i) refine the robot application system according to new product requirements.

Computer simulations are very important in a feasibility study, and such evaluation of the design will be a shortcut to success. To understand a specific problem area, it is important to make two or three advance attempts. This will help to ensure successful technology for practical use, maintenance and reliability. Meanwhile, various aspects of evolving technology can be introduced to the new application. The activities of even the most advanced systems can be improved through continuing experiments.

The next innovation of the robots application systems, with respect to the unmanned FMS, is considered as follows:

(a) approaching FMS by application of the most advanced industrial robots;
(b) designing FMS through application of common industrial robots.

Furthermore, a new integrated production concept becomes available for practical use. This example of an unmanned process consists of a buffer and a robot assembly cell between two in-line production processes. Basically, a small buffer between line A and line B is required to connect the lines. The purpose of the buffer is the smooth handling of works for unmanned manufacturing, and TAKT balance is very important for automated lines. If not balanced, the production line may suffer considerable line disturbance. The control of connecting a few assembly processes can introduce the 'PULL MODE' operation, which is optimum WIP. The whole process is dynamically controlled using the shop-floor control system and a so-called 'IC' card which is a small memory unit for the information controlling the operation. All production information will be retained and used to control the next process. If required, the information will be transferred to a host system and distributed to people required to take quick actions. The assembly production line is fully automated by the computers: this is computer-integrated assembly (Figure 3.12). Normally the buffer sizes are optimized by computer simulation of the production process. Recently, cell-type production has been applied to specific processes in mixed production in Japanese enterprises.

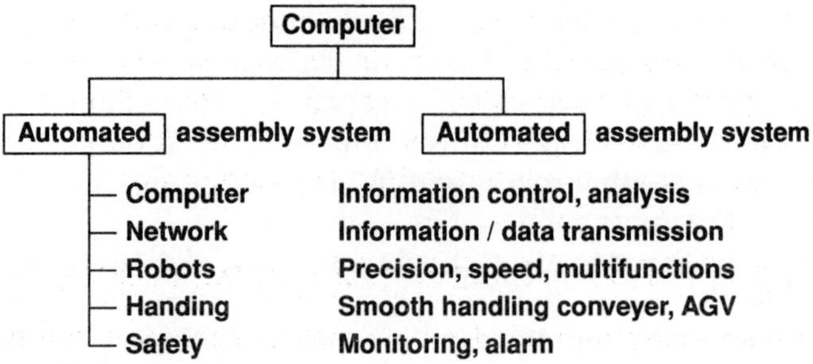


Figure 3.12 Computer-integrated assembly.

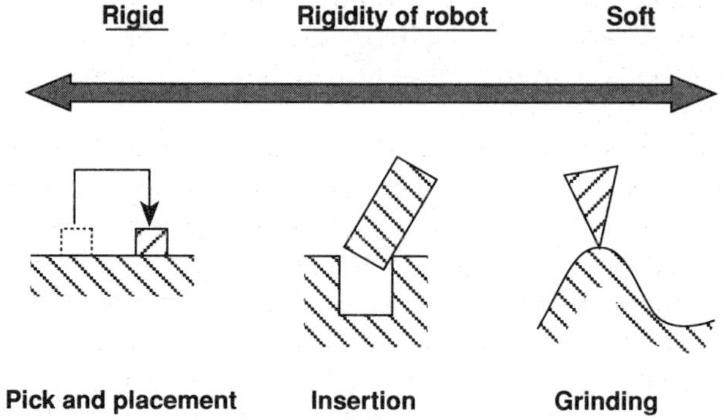


Figure 3.13 Compliance of robot.

Most assembly robots are used for pick-and-place jobs instead of manipulative operations. In order to provide repeatability, manipulators should be rigid, and most industrial robots provide very stiff motion, which is not suitable for insertion operations. To achieve the flexibility of the human hand, IBM Japan has developed a compliance-controlled soft motion robot (SMR) that can move delicately enough to perform insertion. It has six degrees of freedom, allowing it to move freely in three dimensions in order to hold, lift and carry objects (Figure 3.13).

When the pick-and-place robot is used in precision assembly work, an object in the robot's hand is occasionally crushed or broken. To prevent such losses, simple physical measures have been used consisting of a force sensor with microcomputers. The sensor detects the counter-pressure on touching an object. Then high-speed calculation processes the resistance force and controls and robot hand to move on a parametrically defined basis. This cell system for insertion jobs consists of a microrobot and compliance motion controller, a vertical articulated robot manipulator with six degrees of freedom, and an IBM PS/55 computer. Conceptually, the force power of the movement is automatically adjusted and the object is seated by detecting the position where a pin connector is inserted into an adaptor board.

### 3.4.4 LABOUR SAFETY REGARDING INDUSTRIAL ROBOTS

There are so many industrial robots installed in the factories of large enterprises that safety is an important consideration when a worker inspects a robot. It is ironic that the industrial robot, originally introduced to avoid having people do dangerous work, can itself cause an accident by worker carelessness. Some labour accidents can be caused by an unusual function of the robot. Protection must, therefore, be applied to avoid an accident when workers have to enter a dangerous area of the robot's working space for teaching, maintenance, inspection or adjustment of accessories.

In practice, a zoning concept should be introduced for the working space of the robot. It is then essential to use a safety fence with danger signs, a safety door with sensors, and an emergency stop switch which cuts off the power. Although a fail-safe design must be built into the robots, constant awareness of personal safety is still the best way to overcome this problem. Danger can also be reduced by moving slowly and wearing a helmet. An even more effective safety step is to establish a safety standard for working, in which two men work with a watchman who has access to an emergency stop switch.

Another aspect of robots is their reliability and maintainability. Reliability of industrial robots is, at present, extremely low in terms of mean time between failures (MTBF), and the reliability technology for robots and their control has not yet been established. It should be recognized that the possibility of a runaway in work done by a robot still exists. Safety systems and safety operation improvements are therefore very important subjects to be considered when planning robots and designing robot systems. The initial safety steps necessarily involve robot manufacturers and system engineering companies. But in practice, it is more important to involve the users whose workers are subject to robot-related troubles. For greater safety, it is essential to devote the necessary time through education and training in the safe use of robots and automated production.

Considering the popularization of industrial robots in large, medium and small enterprises and their application to night work, the subject of labour safety should be discussed in more detail. One proposal is for a monitoring system that has several functions such as checking the controller, logging of status, indicating problems, analysing problems and communication. In the future, problems will probably be diagnosed automatically and solved easily by a computer-controlled monitoring system using artificial intelligence (AI).

## 3.5 COMPUTER-AIDED QUALITY CONTROL

Quality control (QC) in Japan has traditionally been performed using the standard QC tools, reliability test and inspection. The function of QC is to enable a company to provide products that satisfy customer needs. This function can be efficiently performed by using various data and statistical analysis. Total quality control (TQC) now applies the concepts and techniques of QC to every aspect of corporate activity in Japan. This company-wide TQC requires not only people but also production processes, and these factors are driving effective QC toward the computer-aided QC that is an element of CIM systems. The objectives of CAQC are:

(a) to improve product quality;
(b) to increase productivity in the production process;
(c) to take quick actions on products in the field.

### 3.5.1 QUALITY CONTROL

QC is thus concerned with design quality, inspection and testing of products, and the corrective actions necessary to eliminate poor quality. Statistical quality control (SQC) is currently very popular for improving not only product quality but also job quality. One typical application of control charts in SQC tools derives from the quantitative analysis concept, and control charts are used to stabilize the production process. This is one of the applications to QC of the CIM system. To produce higher-quality products, various testing methods are normally adapted in the final production process. QC in the PCB production process consists of:

(a) inspection of incoming components;
(b) in-process test of logical functions under normal conditions;
(c) functional test of logical functions under normal conditions;
(d) reliability test under simulated stress conditions;
(e) final test in simulated customer environment;
(f) inspection for preventing escapes and for appearance.

The quality of the PCBs can be monitored at each test process; corrective actions will be taken after analysis of all the test results.

### 3.5.2 CAI/CAT

Computer-aided inspection (CAI) and computer-aided test (CAT) are performed automatically using a process computer and sensor technology. Both CAI and CAT, which are a major part of CAQC, are basically stand-alone systems called 'islands of automation'. To improve QC, these systems should be integrated into a host system in which the QC activities are expedited. In the PCB inspection process, for example, it seems unlikely that solder problems such as poor solder and over-solder can be eliminated by functional testing. Soldering should be checked by automated inspection systems instead of by human inspection: that is, by CAI. The result of the inspection is quickly fed back to engineering and the previous production process. QC action can then be taken to improve the production process and to produce superior PCB products without variance.

As an example of CAT, when testing electrical products such as computers, it is essential to perform a 'dynamic test' in order to guarantee reliability and functionality within a certain time. At present it is necessary to speed up clerical information and data processing by automatically collecting and processing assembly testing and inspection results. The highly intensive intelligence resulting can be used to control production schedules, reducing lead time (see Table 3.2 and Figure 3.14). This testing is normally accomplished by go/no-go testing, which quickly determines whether or not a measured value is within specifications. In the measured-variable case of complex products, the value of the quality characteristic is determined by using on-line measuring

Table 3.2 Computer-integrated test

| Process | Information | Control of equipment level |
|---|---|---|
| Preparation for test | Instruction on test location<br>Determination of test conditions<br>Completion of preparation | Mechanical/electrical connection<br>Setting up of various parameters<br>Confirmation of connectability<br>Self-diagnosis of functions |
| Execution of test | Selection of test items<br>Initiation of test<br>Measurement of value | Execution of test programs<br>Data requisition/analysis<br>Report on monitoring conditions |
| Post-process | Completion of test<br>Result of test | Mechanical/electrical disconnection<br>Data transmission to host system |
| Termination | Comprehensive judgement | Integration of materials and information<br>Transfer to next process |

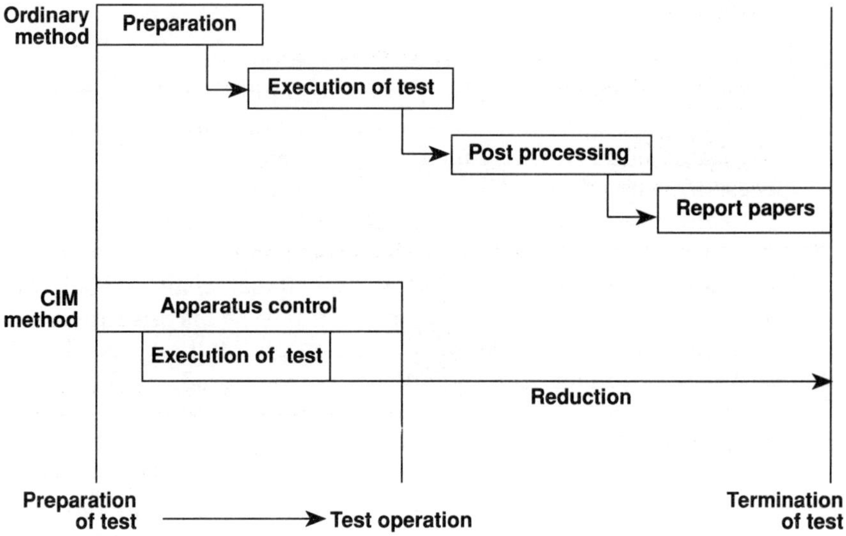


Figure 3.14 Computer-integrated test.

instruments based on an X-chart. Trouble-shooting and optimization to reduce processing time can be based on data from the actual production process, so that the results of the test and data analysis reflect actual operating conditions and parameters. The production of rejects due to missed operations, poor assembly, loose screws, loose connections and other human-related errors can be prevented by using the automation control and on-line monitoring providing by CAT.

CIM, which integrates CAI and CAT, provides highly reliable products and can eliminate many defects in the manufacturing process. Moreover, there are several test processes in the production line, and so integration of all test processes is required to evaluate production as a whole.

### 3.5.3 RELIABILITY IMPROVEMENT

In addition to the use of CAI/CAT, high reliability of Japanese products is also assumed by the traditional method of screening out poor components with stress screening using stand-alone equipment. Production forms and characteristics have recently changed dramatically owing to changes in the economic, social and technological environments. In this situation, personalization and reliability of products are highly appreciated. The traditional manufacturing mentality, which is called 'product-out', should therefore be completely changed to customer-

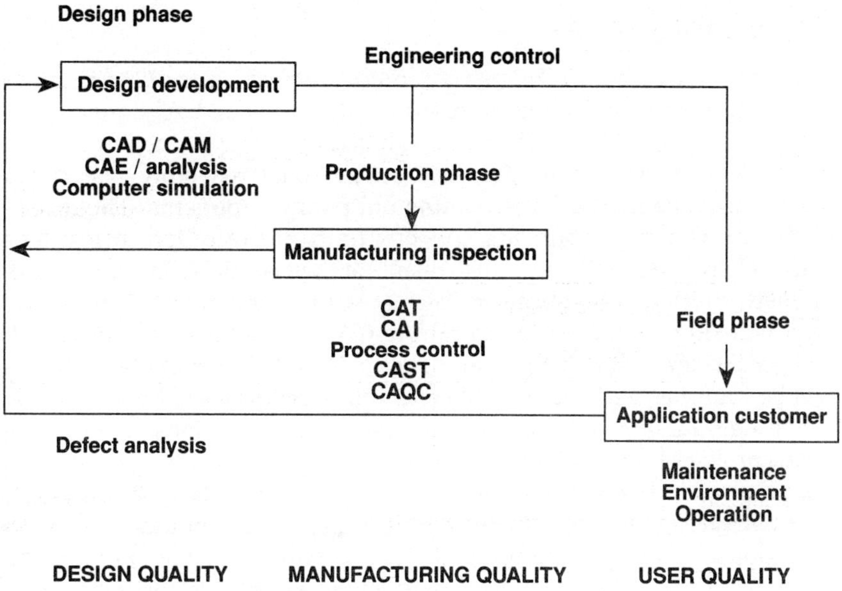


Figure 3.15 Reliability improvement.

driven manufacturing, which is called 'market-in'. To achieve volume production of a variety of different products, it is necessary to use CIM, which is focused on both product reliability and on production efficiency. The core of CIM is total optimization of an enterprise from product design to customers in the field.

The production information for the design phase, production phase and field phase should be turned over as rapidly as possible to improve reliability and eliminate defects and waste (Figure 3.15). In the design phase, it is clear that CAE is needed to improve product design from the QCD point of view. For example, as the result of assemblability evaluation of a new method in the early stage of design, products and assembly process are simplified; this simplification makes it easy to introduce assembly automation and corrective actions for assembly cost reduction.

Various evaluation methods will be much more widely used, just as computer-aided simulation methods already have been. To verify the mechanical stress condition during mechanical design, computer-aided analysis is useful for strength and vibration analysis. When design and production phases overlap, it is difficult to optimize the buffer sizes, conveyor lengths and lot size during actual production. In these cases, computer simulation gives optimized results based on many kinds of simulations in a short time.

### 3.5.4 RELIABILITY TESTS

In addition to CAI/CAT, computer-aided stress tests (CAST) automatically test reliability under stress conditions. The CAST method to remove early failures should be introduced into the production process in order to achieve zero defects. A traditional reliability test called burn-in is just a method of screening out poor components. The results of the CAST for components and box units are recorded in test logs and are updated with new test results either continuously or at predefined intervals. A computer calculates the component degradation forecast using all data collected during CAST, and the calculated result gives a decision on the product reliability. With CAST, quality problems will be found as early as possible and future failures will be forecast by actual degradation data obtained under stress conditions. One of the major problems associated with the stress test is to decide the optimum hours, appropriately balancing the needs of reliability and total cost. The CAST can be used to find the optimum test hours and stress conditions by using experimental design. After a certain period, the stress test time can be reduced gradually by actual test performance.

Regarding QC in the field, most problems can be categorized into two types, infant mortality failure and wear-out failure. Manufacturers are forced to improve quality to minimize the number of these failures. In order to improve field quality, computers can be used for failure analysis to improve both design quality and production process quality. These results should be fed back to the responsible functions or laboratories so that corrective action can be taken as soon as possible. The strategy for CAQC is basically to implement quick feedback between the design, production and field phases.

For manufacturers with highly productive factories, reliability and quality control of the production process, including robots and systems, becomes more and more important. In this case, it is useful to implement a system which monitors such functions as status, condition, problem cause, process time and operation count. The computer-controlled production lines supported by robot monitoring systems will be introduced into the high-grade FMS on unmanned processes.

In Japan, after 15 years of field work, over 180 000 industrial robots have been installed. Now the second generation of the robot age is coming into view. Within a few years almost all industrial robots will have sensors and artificial brains, and the robot's motion will be dynamically controlled enabling their use in high-grade FMS lines. The FMS lines need a highly reliable system combining an extensive range of complex technology (mechanical, electrical, mathematical and computing). In such a system, the reliability and quality control technology of robots and computer systems are key factors for improving manufacturing productivity. Therefore, as the technical concerns in manufacturing engineering increase, Japanese manufacturers will put greater emphasis on manufacturing research and advanced manufacturing engineering. In another important aspect related to FMS, operation and maintenance will be completely changed, so manufacturers must have training and education to expand utilization of the system.

## 3.6 CIM CONSTRUCTION

It is very important that considerable technologies and techniques should be built into the system, as factory construction requires long-term control of the process, progress schedule and budget. Even a small-scale application system for automated processes contains complex mechanical, hydraulic, pneumatic and electrical technologies related to environment control, quality control, inspection, computer software and the human interface. Naturally, as with a construction project, the greatest management skill and the concentration of the most talented staff members and most advanced technologies should be assembled, because their results determine the quality improvement, cost reduction and quick product delivery needed to increase competitiveness. It should be recognized that well-organized technical staff members positively effect the future of an enterprise.

In general, CIM has several targets for improving the productivity of various types of industries:

(a) marketing reinforcement;
(b) product development reinforcement;
(c) business management strength;
(d) quality improvement;
(e) flexibility improvement.

The reader is referred back to Figure 3.5, where CIM for product leverage is described conceptually.

### 3.6.1 CONSIDERATIONS OF CIM CONSTRUCTION

From the user's point of view, many engineers and operators are interested in CIM and want to have a practical system. But up to now, the system has been constructed only by systems engineers; therefore, the following questions arise when the CIM system is constructed. What is the CIM objective? What is the CIM size? What is the area of emphasis? For example, since expansion of business share is assumed as a business strategy, one goal will be to shorten delivery for improving competitiveness. Therefore, it is assumed that the lead time from receiving an order to shipping to the customer should be reduced by 50%. To meet this tough target all functions must take simultaneous action for lead time reduction, such as order processing work in the marketing department, design work in the development department, production process in manufacturing, and planning in production control and purchasing. The target for lead time reduction is distributed individually to each function, then a requirement of the CIM system is defined by the individual target.

Another example is marketing power reinforcement based on data analysis for marketing surveys and customer's information. Business growth is achieved by product planning for customer satisfaction.

To understand various customer needs, a selling agent deals not only with company A's products but also with the products of companies B and C. Based on the analysis of data collected through a computer network and the combined information on A, B and C, superior products can be offered to customers in time.

The production of various products, including small and large, low and high volume products, etc., is integrated and placed under the CIM systems to realize the objectives of the enterprise. It is essential that the CIM system is constructed with attention to the wide view and is implemented in steady steps. For example, IBM CIM-CP is a useful method for achieving the strategic business target and establishing the project development plan.

### 3.6.2 TOP–DOWN APPROACH

The construction of an efficient CIM system is a thoroughly investigated subject. A conceptual construction of CIM results from the combination of three approaches: top–down, prioritized-step and as-required.

The business structure of the enterprise should be established by a business target strategy, and the construction of CIM based on this business target strategy is a key to success. CIM must be an innovative and strategic information system reflecting the business strategy of the enterprise. This approach is a target-driven method for achieving CIM, and is effective in the competitive market. Sometimes, both the top–down and bottom–up approaches are combined to create successful systems from a long-range view. To define clearly the objective related to the business target of the enterprise, top executives should drive CIM construction.

### 3.6.3 PRIORITIZED-STEP APPROACH

It is expensive to construct a system, and a project team composed of technical staff must make a project plan that includes functions, schedules, estimated cost, and effectiveness from the business viewpoint. The selection of various technologies will probably be adapted by justification bases or for easy implementation of each product project. Possible restructuring to solve problems and accommodate new products should be considered to determine the optimum investment. This approach aims at effective implementation during low business growth, and future enhancements will be implemented in prioritized steps. As far as CIM system evaluation is concerned, certain measurements can check the performance or be used to find out how each system's users would evaluate the CIM.

### 3.6.4 AS-REQUIRED APPROACH

This approach is a conservative one in which the system is developed on an as-required basis. This approach aims for effective construction in the short term. The construction method is schedule-driven step implementation, which is instigated by those directly concerned. Each individual system installed on an as-required basis will eventually be integrated. Normally, a master schedule plan is defined by the individual step schedules, and the details of each step's activities are defined when starting.

The project construction steps below are a common approach:

| *Step* | *Technologies and activities* |
|---|---|
| (1) Basic planning | Development plan, alternative investigation, cost estimation |
| (2) Concept design | Function, architecture, layout, logistics |
| (3) Detail design | System requirements, function analysis, application analysis |
| (4) Implementation | System test, reliability test, acceptance test |
| (5) Maintenance | Improvement, maintenance, training |

With the bottom–up approach, implementation is reliable and many people are involved. Future enhancements and modifications, though, will be required.

## 3.7 FUTURE IMPLEMENTATION

As the 21st century approaches, many people are discussing the future of the Japanese manufacturing industries. A committee on the Development Industry Association, headed by Prof Furukawa of the Tokyo Metropolitan University, published a report on future factory systems in May 1989. Also, many enterprises in Japan are just beginning to

Table 3.3 'Road map' of manufacturing technology

| | Year | | | |
|---|---|---|---|---|
| | 1970 | 1980 | 1990 | 2000 |
| Growth | High | Steady | Reasonable | Flat |
| Productivity | Direct | Indirect | Business | Global |
| Product | High volume<br>Low cost<br>Displaced labor | Diversification<br>High quality<br>Improved capacity | Personalization<br>High reliability<br>Higher speed | Customization<br>Robustness<br>Humanistic |
| Form | Automation<br>Robot | FMS<br>CAD/CAM | CIM<br>Network | FCIM<br>Multi-media |
| Computer | Host | Micro/personal | Workstation | Super PC |
| System | Stand-alone | Island | Integration | Synergism |
| Outcome | Products | Service | Customer satisfaction | Paradigm |

consider future business trends, and they are going to establish their plans to cope with the future's demanding business environment. It is essential that research and development activities for future systems collaborate with a university, especially in Japan, where CIM is a focus for university curricula. Looking at the current environment, several problems are raised by the process of CIM implementation:

(a) The system becomes very large.
(b) More advanced technology is required at reasonable cost.
(c) The production increases processes and changes, in the assembly line.
(d) Product quality must be guaranteed in production.
(e) The work environment must be made more comfortable.

Future prospects for CIM are greatly enhanced by new technologies in computers, advanced semiconductors, optics, and associated software. CIM will provide greater flexibility in production systems to deal with a variety of different products in the near future, and unmanned factories will probably become feasible within a few years. In the long term, many industries have to each have strategic CIM which is 'customer-driven' manufacturing to achieve **highly productive and high-value-added** factories. The 'road map' of manufacturing technology is described in Table 3.3 as a typical technology trend. Looking toward the 21st century, the future CIM system is a key to customer-driven manufacturing and to realizing human society.

# Safety in the CIM factory 4

*Noboru Sugimoto*

Safety problems centring on mechatronic equipment (computer-controlled equipment) have recently become topics of discussion. Since these problems largely arise from the user's excessive reliance on computer control, it is not easy to design a computer-controlled machine in such a way that the manufacturer can assume full responsibility for the safety of its use. As the computer structure itself has many unresolved safety problems, machine users are irresponsible if they think that it is a simple matter to have machine control entirely performed by computers. They, therefore, compound the problem by being rather tolerant of this 'computer myth'. Computer-controlled equipment must be designed so that, if it fails, it fails on the safe side and never on the hazardous side.

## 4.1 SAFETY TECHNOLOGY INNOVATION

Today, whether we like it or not, in our society we are forced to live with technological innovation in all fields from industry to household. If we seek higher efficiency or greater ease too impatiently, however, we may learn a hard lesson: technology is handled by human beings, and human habits do not change as rapidly as technological innovation does. Accidents, even disasters, caused by this lag between the introduction of novel technology and the user's learning the preferred habits for its safe use, are a serious social problem today.

Advanced nations have introduced the concept of product liability (PL), placing the fundamental responsibility for such accidents on the

*Manufacturing Automation Systems and CIM Factories*. Edited by K. Asai, S. Takashima and P.R. Edwards. Published in 1994 by Chapman & Hall, London. ISBN 0 412 48230 4

product manufacturer. And there has been a strong tendency to deal with the risk of accidents in the use of computer-controlled machines exclusively through this PL concept.

In 1989, $50 billion was paid out in 17 000 PL cases in the USA alone. Despite the enormous growth of this problem, the USA has not yet found a technical rather than legal solution. This is a serious and difficult situation because it fails to take into account the nature of today's innovation.

Previously, solutions to the safety problems associated with technological innovation could reasonably be expected from the technology itself. Today, though, safety is required from innovative technology that is inherently uncertain. If a technological innovation is needed in order to introduce safety to the use of new and old technology, it must come from the innovation of safety technology. This, realized in fail-safe technology, has recently originated from Japan.

The world of fail-safe technology makes no unreasonable requests, neither for trouble-free machines nor for error-free human beings. Instead, the fail-safe technique assumes mechanical troubles or human errors as a precondition. To prevent such troubles and errors from causing accidents, a safety control system based on safety logic is built into the mechanical equipment. As the requirements of such technological systems have become clearer, fail-safe sensors, logical elements, mechanical elements and other basic devices applicable to safety-oriented systems have been developed.

The conventional method in which the probability of accidents is calculated and one decides whether or not this probability is allowable is no longer adequate. The concept of PL was not designed simply to expose insurance companies to lawsuits, but rather to ensure the production of goods that can be used safely. This is the very 'safety technology' the present authors have been seeking. This chapter will describe safety technology as an original technology independent of conventional high-efficiency and high-amenity technology. It will also show that safety technology is based on fail-safe technology. These points will be demonstrated with an actual manufacturing method using the safety system based on this safety technology.

## 4.2 COMPOSITION OF THE SAFETY CONTROL SYSTEM

### 4.2.1 STATE OF UNCERTAINTY

Between two well-defined operational states of a machine, the state in which safety is confirmed and that in which hazard is confirmed, there is another state which cannot be determined as being either safe or hazardous. In this state, it is uncertain whether or not any safety

Table 4.1 Judgment on continuation of operation

| Insufficient safety | | Sufficient safety |
|---|---|---|
| Continue operation | Safe state | Continue operation |
| | Uncertain state | |
| Discontinue operation | Dangerous state | Discontinue operation |

device can detect danger and function properly until the machine is actually operated.

Operating a machine in such an uncertain state is strictly prohibited because it should be regarded as putting human safety at risk. According to the safety system proposed by the Research Institute of Industrial Safety (Ministry of Labour of Japan), the system should be allowed to operate only when the safe state is confirmed (see Table 4.1).

### 4.2.2 HAZARD DETECTION VERSUS SAFETY CONFIRMATION

The safety confirmation system, which is the logical basis of fail-safe technology, has been employed in the production lines of large Japanese companies such as Bridgestone, Kao and Asahi Chemistry.

Safety confirmation devices, or sensors for safety detection, play important roles in all safety systems. Various kinds of detectors are used to detect 'safety'. A typical example in common factories is a reflection-type photoelectric switch that detects humans approaching the specified hazardous zone. Another example is a limit switch used at various places. Many of these switches are not activated as long as the master machine works normally, but whenever the normal state turns hazardous, these switches detect the hazard and warn of the danger or stop the machine.

If this type of detector is out of order, however, the hazard may not be detected. For example, the light receptor of the photoelectric switch in Table 4.2 is in the OFF position during a normal safe state, but it is not certain whether the OFF position means that the machine is actually in the safe state or simply that the switch is out of order. This is the hazard-detection type of system. It makes an unreasonable request to the safety device: that even though the safety device may someday be out of order, it must function normally whenever a hazard exists.

In the safety confirmation type of system, the machine always starts or continues operation based on information that the machine is in a safe state. In the above example of the photoelectric switch, when the transmission type switch shown in Table 4.2 is in the normal safe state,

**Table 4.2** Examples of light-beam sensor and ultrasonic sensor

| Mode | | Schematic illustration | Description |
|---|---|---|---|
| Photoelectric switch transmission type | Safety confirmation type | 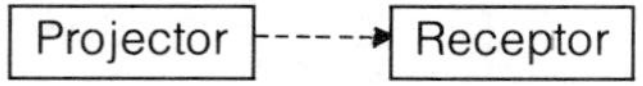 | This mode is used for sensing an object in the danger zone. If there is a failure in the system (e.g. burnout of the light projector bulb, etc.) system operation stops (safety side). |
| Photoelectric switch of reflection type (A) | Hazard detection type | 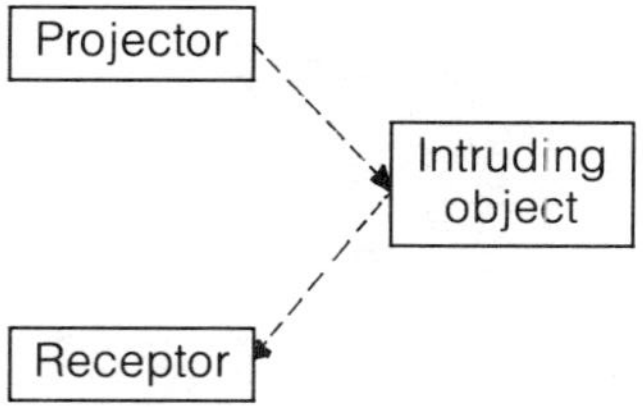 | This mode is used for sensing an object in the hazard zone or for having the unmanned carrier confirm the safety of its forward path. However, if there is a failure in the system (e.g. burnout of the light projector bulb, etc. or if the projector lens is clouded) this mode will not have the system stop operation as a safety device. |
| Photoelectric switch of reflection type (B) | Safety confirmation type | 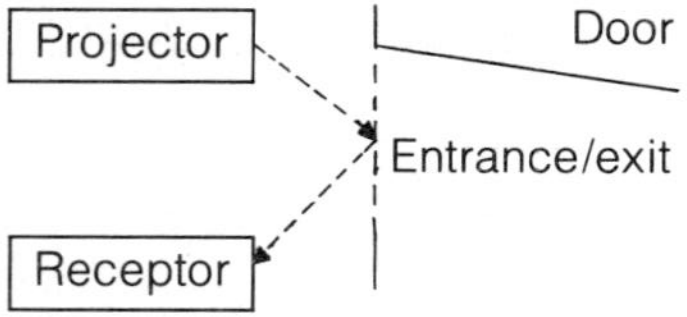 | This mode is used for confirming the closed state of the door. If there is a failure in the system (e.g. burnout of the projector bulb) this mode will have the system stop upon opening of the door (Can be used as a safety device in this way unlike the reflection type A). |
| Fail-safe ultrasonic sensor | Safety confirmation type | 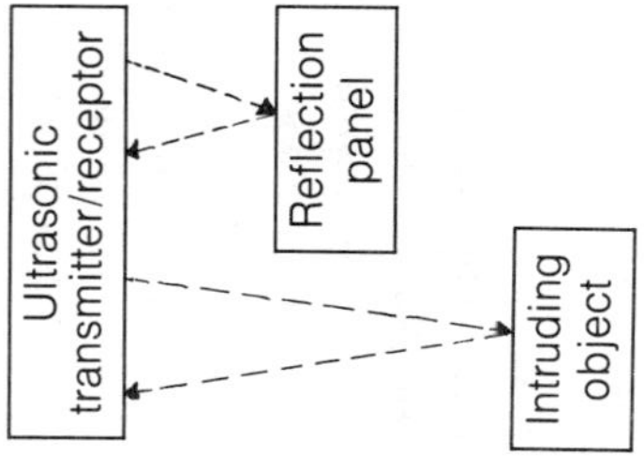 | This mode is used for having the unmanned carrier confirm the safety of its forward path. While keeping confirmation of its own normal function using the reflection panel installed near the transmitter, this type senses the object that is intruding ahead of the sensor. |

light enters the light receptor and a safe signal is repetitively issued. If a human body shuts off the light (hazard) or the light does not reach the receptor owing to some trouble (uncertainty), the machine is stopped by the absence of the 'safe' signal. The system must therefore detect a signal strong enough not to allow the 'safe' signal to be generated by noise and falsely transmitted to the machine in a hazardous state.

## 4.3 BASIC CONCEPT OF SAFETY CONFIRMATION TECHNOLOGY

### 4.3.1 SAFETY CONDITION IN CIM SYSTEMS

There is no doubt that the introduction of CIM systems brings remarkable progress in automation, robotization, computerization, and results in labour saving. A high degree of automation, however, does not necessarily mean increased safety even though, as manual jobs are mechanized and robotized, dangerous moving parts are removed from human control. Many accidents associated with automation occur when workers ignore dangerous moving parts that are not the objects of work during non-steady-state jobs, such as maintenance and troubleshooting. Troubleshooting workers entering highly automated production systems must work surrounded by many dangerous moving parts (see Figure 4.1). In such complex automation production systems, the number of moving parts has so increased that they elude the full attention of even the most careful workers.

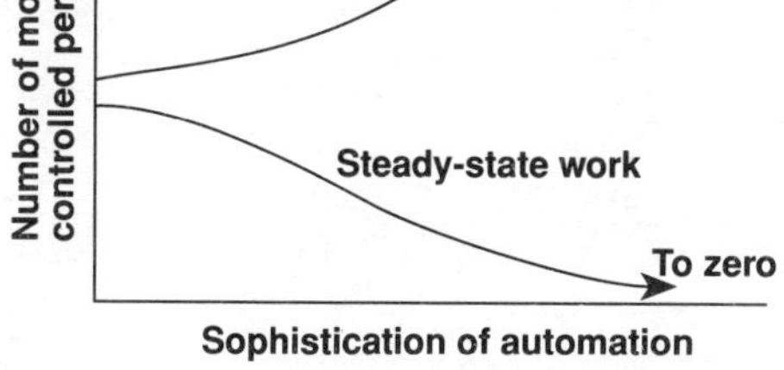


Figure 4.1 As automation advances, the number of movable parts to be controlled increases (after Sugimoto, 1987).

According to safety confirmation theory, the characteristic of extended hazardous spaces around machines follows the process of automation, so that safe spaces for human operators are reduced. The safety of CIM systems should be designed with the consideration that accidents occur due to interference between a human (human workspace: safe space) and a machine (a machine operation space: hazardous space generated by a machine's dangerous energy).

Whenever a machine is operated (or issues energy) in a hazardous space, humans should be strictly prohibited from entering that space. If necessary, in a limited time conditioned by safety (or the energy condition of the machine) humans can be allowed to enter the hazardous space; and if the system detects a human entering hazardous space, it should shut down all machine power.

### 4.3.2 FUNDAMENTAL CONSTRUCTION OF INTERLOCKS

If the factory automation system consists of $n$ unit cells (see Figure 4.2), the movable space of the machine is defined as $x_i$ ($i = 1, 2, \ldots, n$) for the $i$th cell. Let $u(x_i, t) = 0$ be 'stop' information for the $i$th cell at time. Let $w(x_i, t) = 0$ be 'human' absent' information on the space $x_i$ at time $t$ and $w(x_i, t) = 1$ be 'human present' information on space $x_i$ at time $t$. Then the fundamental safe condition is given by the logical product:

$$u(x_i, t)w(x_i, t) = 0 \tag{4.1}$$

In the safety-confirmation system, the machine can be operated only when humans are absent from the space $x_i$ ($i = 1, 2, \ldots, n$). The permission of the machine operation $p$ $(x_i, t)$ is given by a safety device: i.e. a safety-confirmation sensor signalling safety of human absence. The sensor $I_M$ can issue the permission signal $p(x_i, t) = 1$ to the machine.

The sensor $I_M$ should be so constructed that it detects safety and issues safety. That means that signals generated by $I_M$ (safety sensor in general) should be inversely processed. Machine execution occurs as a

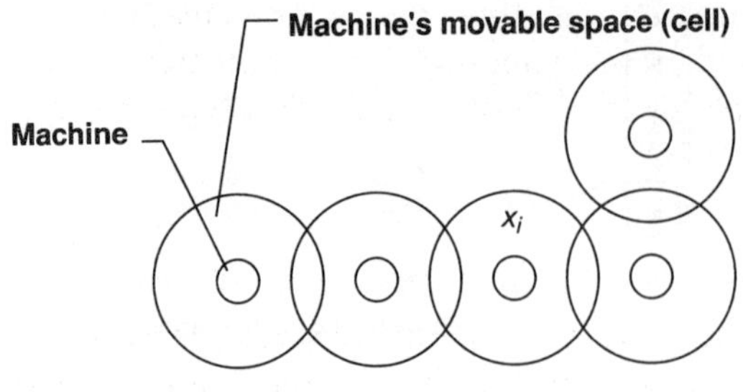


Figure 4.2 FA system in production factory.

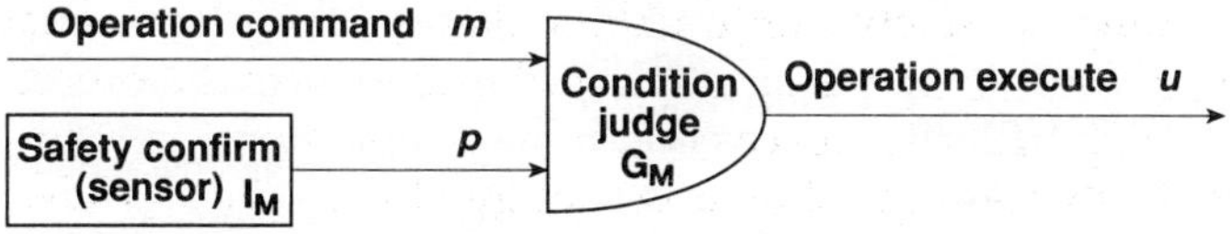


Figure 4.3 Interlock of safety-confirmation system.

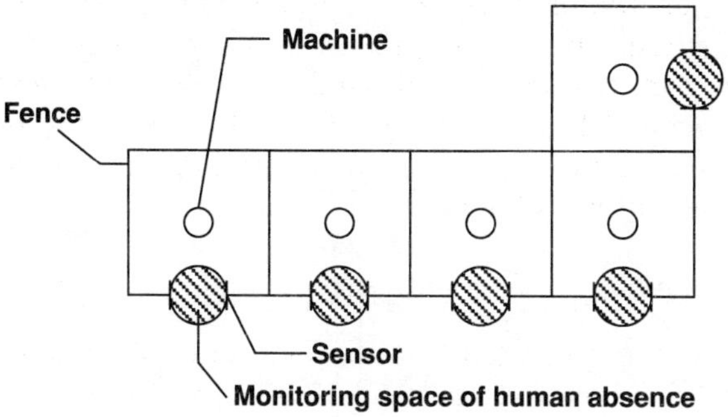


Figure 4.4 Configuration of machine cell sensor.

result of the logical product of operation command $m$ to the machine controller and the permission given by $I_M$.

The architecture to guarantee equation 4.1 is the 'interlock' shown in Figure 4.3, which requires the following conditions:

(a) A 'safe' signal should not be issued when any malfunction occurs in the sensor $I_M$.
(b) When the condition judgment element $G_M$ (if used) is out of order, a signal of 'execution' should not issued.

An example of sensor configuration is shown in Figure 4.4 for a safe work system considering the system of machines shown in Figure 4.2. The machine cell is fenced and has a defined entrance through which humans can enter into the cell space. As shown in Figure 4.3, the interlock is considered in the case of non-steady-state jobs such as maintenance and troubleshooting. The photoelectric sensor in Figure 4.5 is an example of the safety confirmation type that confirms human absence.

### 4.3.3 FAIL SAFE

Mechanical equipment such as industrial robots should be built in such a way that, if any element that is to transmit a safety signal by means of the parts composing the circuit or the device (e.g. electronic parts

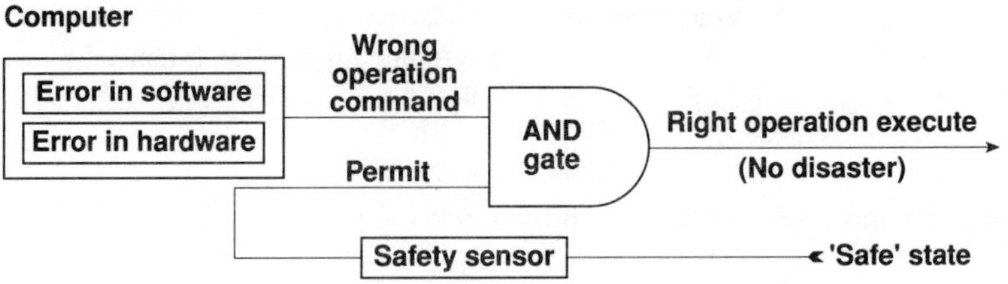


Figure 4.5 Application of interlock to automatic machines.

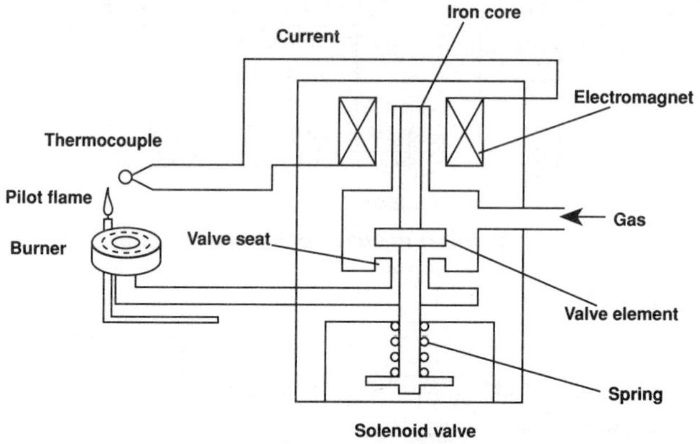


Figure 4.6 Safe gas-supply system in domestic gas hot-water heater.

and machine parts) should fail, dangerous mechanical output stops. The technology designed to realize this function is fail-safe technology.

To clarify the meaning of fail-safe technology, Figure 4.6 shows a specific example of a safety system: a safe gas supply system for domestic water heaters, which can be used to illustrate the fundamental construction of fail-safe technology. The button which opens the valve which supplies gas to the burner simultaneously activates an ignitor. The gas flame induces an electric current in a thermocouple which acts as a simple safety sensor. This current energizes an electromagnet which withdraws an armature and keeps the valve open even when the button is released. In the event of a failed ignition or the flame going out, the electric current will not be induced in the thermocouple. This switches off the electromagnet and closes the valve, shutting off the gas supply. Figure 4.7 shows the relationship between the components of the system shown in Figure 4.6 in terms of working correctly, being in a normal state, and signalling the presence of a flame. Table 4.3 shows the potentially dangerous failure mechanisms of each component. Note that the structural parts allow the flow of gas in

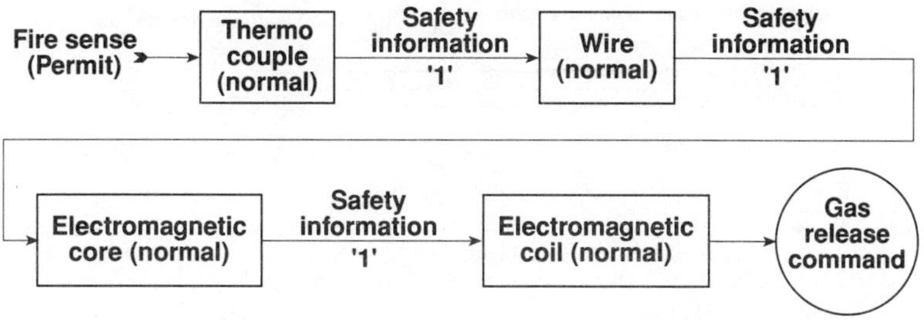


Figure 4.7 Transmission of safety information in gas hot-water heater.

Table 4.3 Analysis of fault mode

| Component | Fault mode | Symptom |
|---|---|---|
| Structural part | | |
| Burner | Clogging | No gas feed |
| Pipe | Crack | Gas leakage |
| Gas chamber | Fracture, hole | Gas leakage |
| Valve | Breakage, bend | Gas leakage |
| Control part (Mechanical) | | |
| Knob | Detachment | No gas feed |
| Shaft | Breakage | Gas leakage |
| Electromagnetic iron core | Detachment | Electromagnet not actuated |
| Spring | Breakage | Gas leakage |
| (Electrical) | | |
| Thermocouple | Breakage | Electromagnet not actuated |
| Wire | Disconnection | Electromagnet not actuated |
| Electromagnetic coil | Disconnection | Electromagnet not actuated |

the event of failure and should be solidly constructed. The control parts are, however, more complex and hence more likely to fail. This examples shows that fail-safe technology is the method of putting together the components of a safety system such that an unavoidable failure of any component of the system results in a safe output from the system as a whole.

## 4.4 OPERATION OF A SAFETY-CONFIRMATION SYSTEM IN AN AUTOMATED FACTORY LINE

### 4.4.1 SEALED BEAM SENSOR

A 'sealed beam sensor' is formed by connecting a light projector and a light receptor with a rubber tube through which light can pass

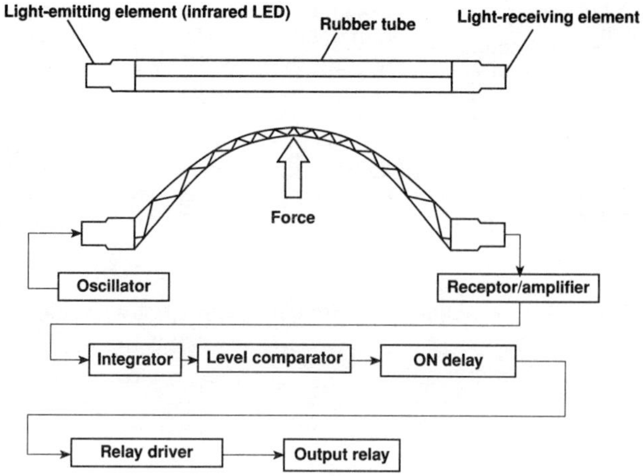


Figure 4.8 Principle of sealed-beam sensor.

(Figure 4.8). Light from the projector passes along the tube with irregular reflection to reach the light sensor. The presence of light at the sensor causes an ON signal (a safety signal) to be issued. Whenever the tube is disturbed, for example by a person, the light beam will be attenuated according to the extent to which the tube is bent. Once the light level at the sensor falls below a threshold value, an OFF signal (a hazard signal) is issued. Furthermore, in the event of a break in the rubber tube, or the failure of the light projector or sensor, an OFF signal will also be issued. Thus the sensor has a fail-safe mechanism. This system has several features which are not realized in conventional light beam sensors:

(a) the lenses of both the light projector and sensor are protected from contamination;
(b) the optical axes can easily be aligned;
(c) accidental contact between personnel and the sensor is reduced due to the visible nature of the connecting tube;
(d) the flexible nature of the tube means that it does not introduce additional injury hazards.

This system can also be used to detect collisions with unmanned carriers, and as a safety fence.

### 4.4.2 THE FAIL-SAFE AND GATE

An AND gate only turns its output signal, $V_0$, on when both of its input signals $I_1$ and $I_2$ are on. A fail-safe AND gate should have operating

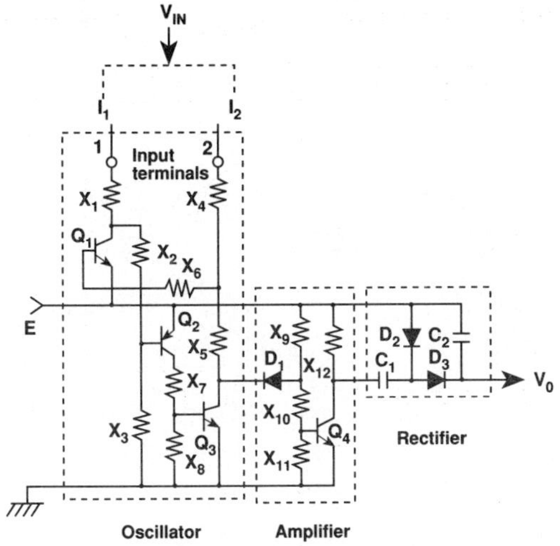


Figure 4.9 Circuit of AND operation oscillator.

characteristics such that the output signal is turned off in the event of any problem occurring. This can be achieved electronically using an oscillator circuit. An input signal from an oscillator needs to be of a higher energy than a corresponding DC signal. The principle of operation is that a faulty oscillator circuit cannot generate high-energy AC signals.

A practical fail-safe AND gate is shown in Figure 4.9. In the figure, OSC is an oscillator; AMP is an amplifier; REC is a rectifier. When the inputs $I_1$ and $I_2$ are not applied to OSC, the output state is:

$Q_1$ OFF, $Q_2$ ON, $Q_3$ ON.

When both inputs $I_1$ and $I_2$ are applied to OSC, oscillation starts in the following sequence:

$Q_2$ OFF $\rightarrow$ $Q_3$ OFF $\rightarrow$ $Q_1$ ON $\rightarrow$ $Q_2$ ON $\rightarrow$ $Q_3$ ON
$\rightarrow$ $Q_1$ OFF $\rightarrow$ $Q_2$ OFF.

The OSC output is generated as shown in Figure 4.10. This output is amplified and then subjected to voltage doubling rectification. This prevents a short circuit from the power supply from generating an output. Table 4.4 shows an analysis of the fail-safe AND gate's potential failure mechanisms. It is clearly shown that this device is guaranteed not to generate any output as the result of a fault.

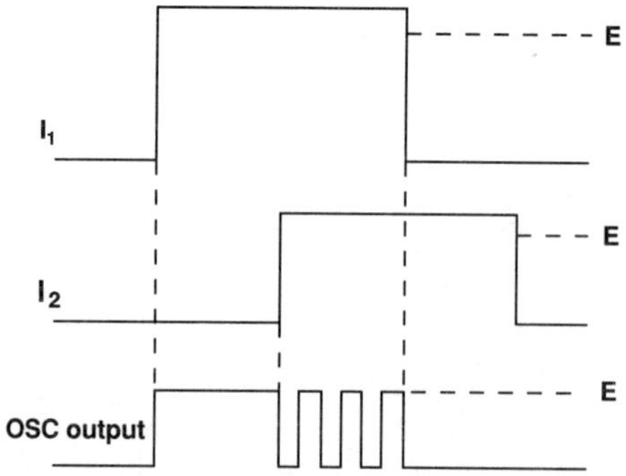


**Figure 4.10** Input/output relationship of AND operation oscillator.

**Table 4.4** Analysis of faults with fail-safe AND gate

| Component | Symbol | Trouble | Symptom |
|---|---|---|---|
| Transistor | $Q_1$–$Q_3$ | Short-circuit/disconnection between each three termlnal | AND operation oscillator not activated |
| | $Q_4$ | Short-circuit/disconnection between each three terminal | No oscillation output or output drop |
| Diode | $D_1$ | Short-circuit | No oscillation output |
| | | Disconnection | No oscillation output |
| Resistance | $X_1$–$X_8$ | Disconnection | AND operation oscillator not activated |
| | $X_9$–$X_{12}$ | Disconnection | No oscillation output |

### 4.4.3 APPLICATION TO A SEESAW MODE OPERATION OF A ROBOT IN A CELL

Seesaw mode robot operation can be used for tasks such as the arc-welding operation shown in Figure 4.11. The robot's working envelope is divided into a left and a right zone. While the robot is performing an automatic welding operation in one zone, a human is loading or unloading workpieces in the other zone. When both the robot and the human have completed their respective tasks, the operating zones are switched over. This is called 'see-saw' operation mode, because the operator and the robot change their respective operation zones while cooperatively proceeding with their jobs, as if they were playing on a see-saw.

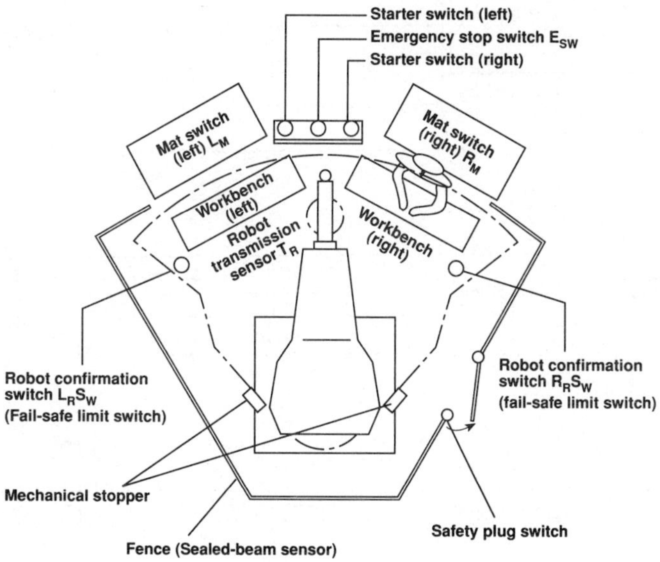


**Figure 4.11** Application of basic safety-control device to see-saw mode.

A safety confirmation system is shown in Figure 4.12 featuring a safety mat, a sealed-beam sensor, a fail-safe position confirmation sensor, a fail-safe AND gate and a fail-safe OR gate (equal to a wired-OR). The operation of this system is as follows:

(a) The robot is in the home position, and the operator presses the right starter button switch. At this time, the left and right mat switches ($L_M$, $R_M$) are effective.
(b) When the robot comes into the right zone and presses the robot confirmation switch ($R_RS_W$), the left mat switch $L_M$ becomes ineffective, and the operator can perform the job in the left zone and the robot in the right zone.
(c) When both the operator and the robot complete their respective jobs, the operator comes out of the left zone and presses the left starter button switch. This causes the robot to move to the left zone. When the robot passes through the arm transmission sensor $T_R$ on the way, both the left and right mat switches $L_M$ and $R_M$ become effective.
(d) When the robot comes into the left zone and presses the robot confirmation switch ($L_RS_W$), the right mat switch $R_M$ becomes ineffective. This enables the man to perform the job in the right zone and the robot in the left zone. When both the robot and the man complete their respective jobs, the man comes out of the right

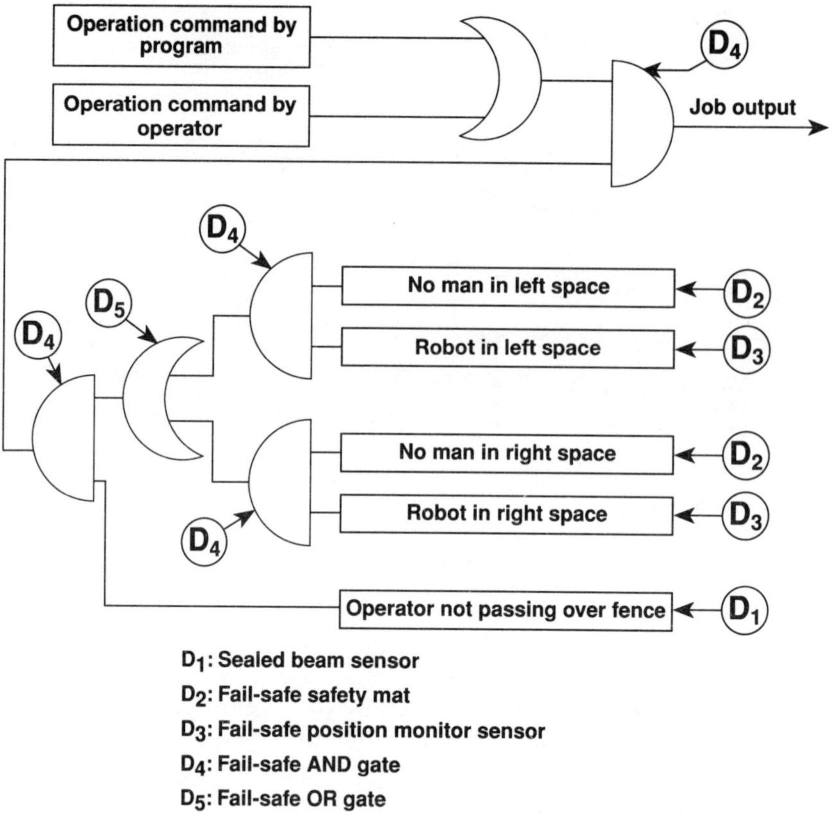


Figure 4.12 Control system for robot see-saw mode. $D_1$, sealed-beam sensor; $D_2$, fail-safe safety mat; $D_3$, fail-safe position monitor sensor; $D_4$, fail-safe AND gate; $D_5$, fail-safe OR gate.

zone and presses the right starter button switch. This causes the robot to move to the right zone. When the robot passes through the arm transmission sensor $T_R$ on the way, the left and right mat switches $L_M$ and $R_M$ become effective. Then the system goes back to state (b) above.

For the time chart, see Figure 4.13.

#### 4.4.4 APPLICATION TO HOLD SUPERVISION

For the mode described in Figure 4.11, there is a case where the robot is checked, repaired or adjusted in a hold-stop state. In this case, the safety of the operator must be guaranteed. A mechanical window switch (shown in Figure 4.14) has been developed for this purpose.

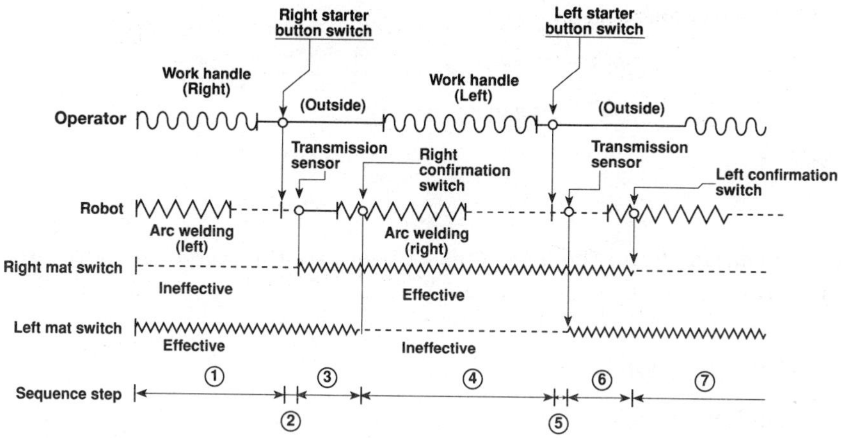


Figure 4.13 Time chart for the safety confirmation system shown opposite.

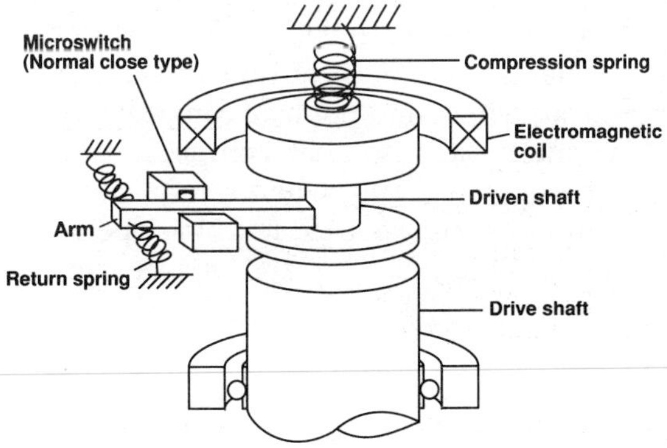


Figure 4.14 Mechanical window switch to help guarantee operator safety.

When this system is in normal operation, the electromagnetic coil is supplied with electric current, and the drive shaft (on the actuator side) and the driven shaft (on the sensor side) are separated from each other by the magnetic field generated by the current applied to the electromagnetic coil.

When an operator comes into moveable range of the robot, the hold-set button is pressed. This interrupts the electric current through the solenoid, and the drive shaft and the driven shaft are mechanically connected (closely attached). This is called the 'hold-stop' state.

If nothing goes wrong at this stage, both the drive shaft and the driven shaft stay put, and the operator can perform his job within the moveable range of the robot. If, however, the actuator should run away due to trouble or noise, the drive shaft will start to rotate, as will

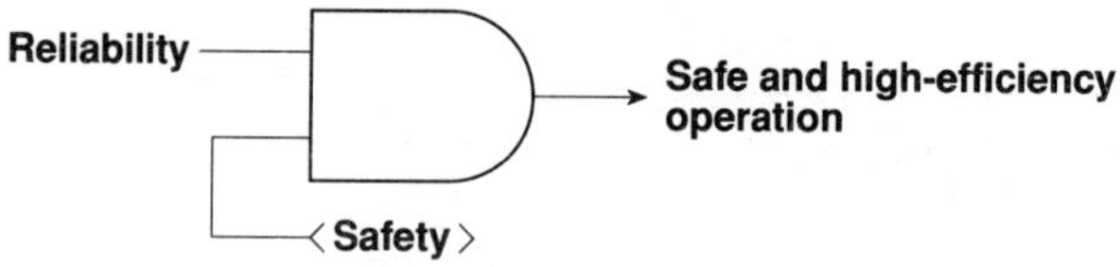


Figure 4.15

the driven shaft and the arm. This will activate the microswitch to turn the drive power source OFF.

## 4.5 CONCLUSIONS

The guiding principle by which safety confirmation systems are designed is: do not take chances unless safety is fully assured. Fail-safe technology incorporates this principle in the control of machinery. When dealing with safety assessment methods, safe working rules, safety management and safety regulations on CIM systems, as well as with technical problems, assurance of human safety must be assured by a proper interlock system, and not by a computer system. Computer systems must be interlocked so that computer-controlled machines and systems can be operated only when safety is assured by fail-safe technology. As shown in Figure 4.15, machine operation systems can then be operated with high reliability and perfect safety.

## REFERENCES

Sugimoto, N. and Futsuhara, K. (1987) Safety and technology of safety confirmation type: coordination with robots. *Proc. Trends in Ergonomics/Human Factors IV*, Miami.

Sugimoto, N. (1985) Subjects and problems of robot safety technology. *Proc. Fifth UOEH International Symposium*, Kitakyushu.

# PART TWO
# Examples and Effects

# CIM systems in the machinery industry

# 5

*Shigeo Hotta*

Recent trends in the changing economic environment of manufacturing industries are toward diversification of market needs, more rapid delivery schedules, reduction of working time and a higher international exchange value for Japanese currency. A major challenge facing industries in this era is the transition from local automation of various individual sections to comprehensive computer-integrated manufacturing (CIM) systems embracing company-wide operations.

The central concept in CIM, rationalization of information, includes three objectives:

(a) synchronizing the flow of articles and information, and facilitating the flow of information by standardization;
(b) establishing channels for prompt and detailed transmission of information;
(c) unifying databases.

These goals must be achieved in order to ensure efficient coordination of information exchange.

Advances in data-handling techniques, with the concomitant cost reduction for information devices and computers, as well as other improvements in operating environments, have already facilitated the rationalization of information. Thus, CIM is no longer a futuristic subject, but now poses an immediate and realistic challenge.

Indispensable factors in this information-oriented trend are intelligent terminals and the technology of networks linking them. The uncoordinated introduction of proprietary networks has led to a plethora of

*Manufacturing Automation Systems and CIM Factories*. Edited by K. Asai, S. Takashima and P.R. Edwards. Published in 1994 by Chapman & Hall, London. ISBN 0 412 48230 4

different networks within factories, impeding information exchange. Thus, network standardization and the organization of communication systems as infrastructural components become necessary.

In this context, the General Motors Corporation in 1982 proposed development of manufacturing automation protocols (MAP) for local area networks (LAN) in industrial factories. Standardization has progressed considerably since then, and the present prospects for international standardization of LANs for factory automation appear quite promising.

Since 1985, Toyoda Machine Works has recognized the importance of MAP as an essential technical factor in the implementation of CIM, and has devoted systematic efforts to the practical realization of MAP. This chapter briefly describes examples of practical MAP and discusses subsequent directions in factory automation (FA).

## 5.1 CHARACTERISTIC FEATURES OF MAP

Several reasons for the attention currently being drawn by MAP as the most promising trend in LANs for factory automation can be cited as follows.

### 5.1.1 INTEROPERABILITY

MAP conforms to the criteria for standard communication protocols prescribed by the International Standards Organization (ISO):

(a) the protocol is open, and readily implemented by manufacturers;
(b) the functioning of the protocol when different terminal models are interconnected can be validated by public test centres.

These advantages make this protocol suitable for use in a multivendor environment.

### 5.1.2 REDUCTION OF THE LOAD IMPOSED ON COMMUNICATION SOFTWARE

The ISO has specified several varieties of standard general-purpose software and, by exploiting these, users can minimize the cost of investment in communication software. MMS software is designed with particular consideration for interfaces with factory automation devices such as NC systems and robots, and it possesses outstanding features not provided by other networks such as Ethernet.

### 5.1.3 SIMPLIFICATION OF CABLE ARRANGEMENTS

MAP uses broadband communication channels, enabling simultaneous transmission of different messages (including graphic information), and thereby reducing the number of communication circuits needed. MAP also facilitates signal branching and wiring, thus simplifying expansion and modification of the network.

### 5.1.4 SUITABILITY FOR FACTORY TRUNK NETWORKS

The introduction of CIM requires large-scale networks covering entire factories. MAP permits the incorporation of approximately 10 km of transmission distance and more than 1000 connection nodes, thus meeting the requirements of factory trunk networks.

## 5.2 PRACTICAL REALIZATION OF MAP IN PARTS PRODUCTION PLANTS

### 5.2.1 A MODEL AUTOMATED FACTORY

The increasing diversification of motor vehicle user needs has given rise to a corresponding diversification of product specifications in parts production plants. As a consequence, it has been increasingly necessary to fundamentally modify factory lines conventionally configured as special-purpose machining lines oriented toward mass production.

The plant briefly described here started full-scale operation in the summer of 1987. This factory, the fifth plant among the facilities of Toyoda Machine Works and devoted entirely to the manufacture of motor vehicle parts, constitutes a model factory automation plant with production lines operated chiefly by automated equipment.

A multivendor-oriented information network conforming with MAP is already used in these new production lines, and the organization and prototypical operation of an automated multimodel parts production system based upon **kanban** (just in time) is in progress.

### 5.2.2 CONFIGURATION OF THE MAP NETWORK

The MAP network in this plant uses broadband coaxial cables (transmission rate 10 Mbps) with a total length of approximately 2 km. A tap-off at intervals of about 20 m is established, permitting connection from any part of the mechanical equipment within 30 m, so the network is readily adaptable to layout alteration (Figure 5.1). The network (Figure 5.2) was designed in consideration of the following points:

Figure 5.1 Broadband coaxial cable in the plant.

(a) maximum exploitation of the broadband characteristics;
(b) low cost;
(c) man–machine interfacing;
(d) multivendor organization.

### (a) Exploitation of broadband features

Utilizing the characteristics of the broadband cable which constitutes the backbone of MAP, every effort was devoted to establishing transmission procedures such that various communication media (even those not fully conforming to MAP specifications) could be integrated into these channels.

The cellular level output ports for data and information perform important functions. Hence, the standard protocol was established to provide compatibility with RS232C and RS-422 interfaces. This permitted the use of low-cost hardware and facilitated the construction of the network.

### (b) Man–machine interfacing

CRT displays and printers are generally used as man–machine interfaces in computerized information networks (Figure 5.3). Production plants,

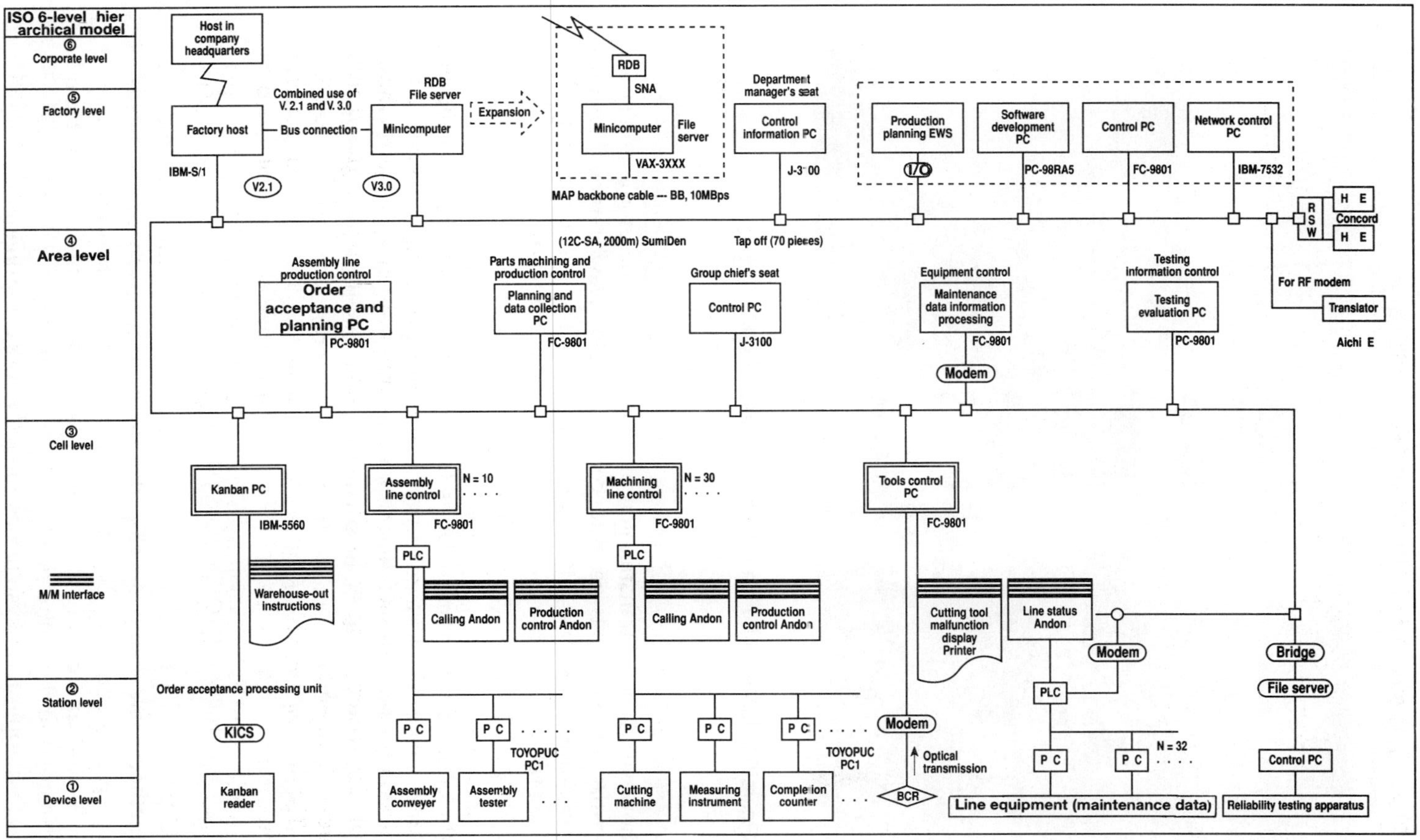


Figure 5.2 Schematic diagram of MAP network.

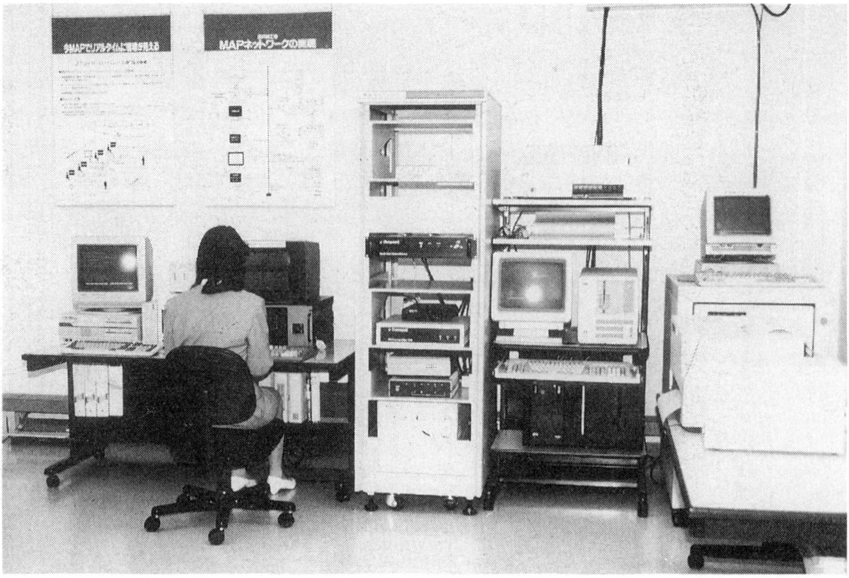

Figure 5.3 MAP centre managing the MAP network.

however, require transmission of information to the production supervisors on simple, large, and conspicuous displays rather than on a multitude of diverse information displays. The system employed by the present company therefore conveys information from personal computers to a PLC, which then displays the information in a large and simple format using an information transmission device called an 'andon' (process indication lamp). Due efforts have thus been devoted to devising lucid means of information transmission (Figure 5.4).

This 'andon' device efficiently transmits information and commands in real time and is thereby conducive to the prompt judgment and decisions needed to make the effective responses that are the practical objectives of CIM systems.

### 5.2.3 FACTORY AUTOMATION ITEMS

Application systems, major items in the improvement of production methods, are being developed with priority given to the needs of the production staff. Based on the prototype system implemented in 1988, the configuration is being revised and modified so that it can be extended to the entire plant.

This application system has the following components:

Figure 5.4 Information transmission device (*andon*).

(a) **Production control system** – production-based sequential order acceptance with the objective of reducing stock to zero in accordance with kanban:
   (i) order placement processing by kanban reader;
   (ii) assembly planning coordinated with kanban reader;
   (iii) relief-ace system (for improved human productivity);
   (iv) 'andon' system (production management and display);
   (v) confidential personnel planning.

(b) **Cost control system** – immediate daily action based on accurate daily information on tools, working days and rates of operation:
   (i) management of basic cutting tool units (daily tallies, monitoring abnormalities);
   (ii) management of basic clerical supply units;
   (iii) management of tool expenditure budget;
   (iv) calculation of manufacturing costs (using host computer at company headquarters).

(c) **Quality control system** – improved quality control by accurate assessment of realtime quality data:
   (i) gathering and management of defective articles in the production process;
   (ii) management of mould control data;

(iii) management of reliability testing data.

(d) **Equipment maintenance control system** – preventive maintenance by prediction and early detection of malfunctions:
   (i) monitoring high-speed spindle lifetimes;
   (ii) detecting early malfunctions in new equipment.

## 5.3 AN EXPERIMENTAL FACTORY AUTOMATION TRAINING SYSTEM CONFORMING WITH MAP

### 5.3.1 INTRODUCTION

The present system was delivered to the Industrial Research Institute that the Aichi Prefectural Government established in 1981 at Kariya for the purpose of stimulating the development of small and medium-scale industries (machinery, metals, chemicals, civil engineering).

This institute was formerly equipped with NC industrial machinery, robots, personal computers and other sophisticated devices and was active in factory automation technical training. As the institute's equipment increased to allow research, development, and practical training related to still more advanced factory automation, a MAP network was subsequently introduced in order to cope with the need for controlling factory automation devices provided by different suppliers.

This experimental factory automation training system is characterized by the following features:

(a) Equipment previously at the technical centre has been incorporated in the system.
(b) With some exceptions, the various items composing the system were supplied by different manufacturers.
(c) Personal computers have been used to the maximum possible extent.
(d) The system possesses a low-cost communication network with 5 Mbps carrier band specifications (V2.2).

All of these features have been provided in consideration of the circumstances of factory automation in small- and medium-scale industries.

This is the first introduction of a MAP-compatible factory automation system at a public research and testing institution in Japan and has attracted considerable attention for this reason.

### 5.3.2 SYSTEM CONFIGURATION

The present system includes stations for machining, inspection, assembly and ASAR (automatic storage and retrieval). An AGV (automatic guided vehicle) travels between these stations, performing

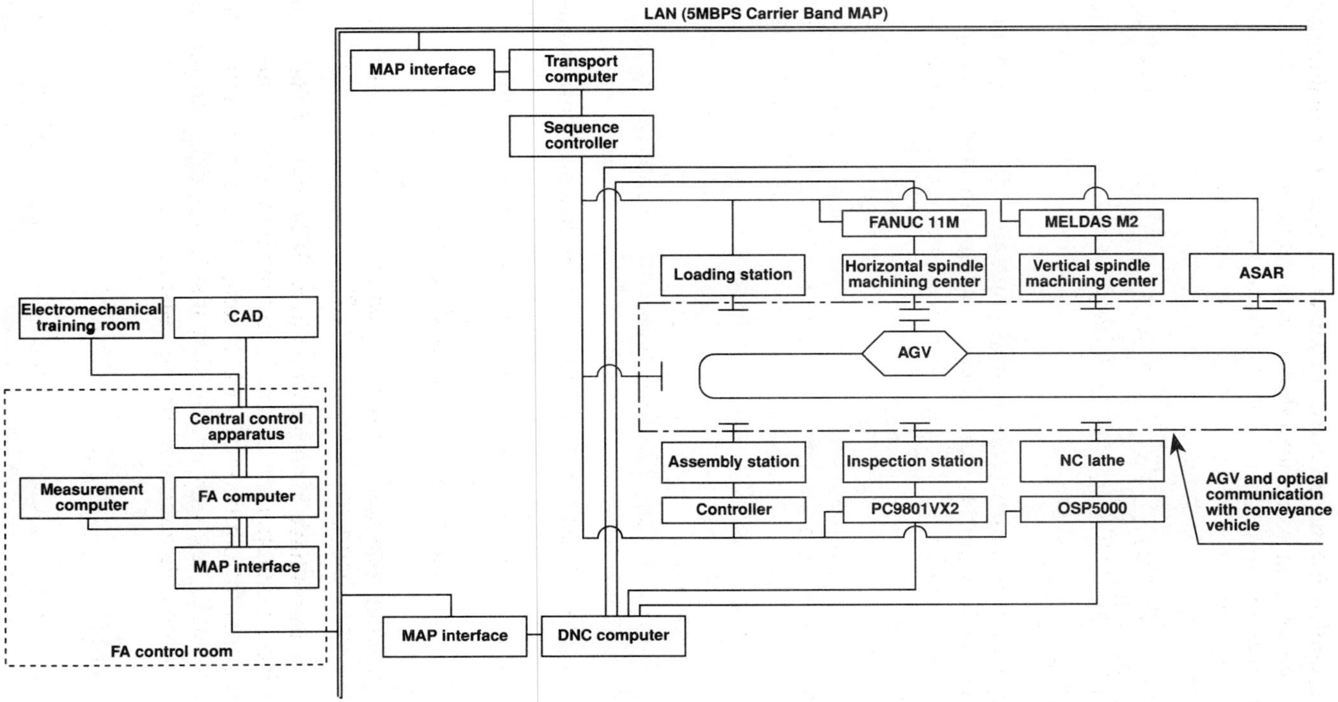


Figure 5.5 Configuration of experimental FA training system.

various tasks – including delivery of materials, machining, and product output – guided by the central computer in the factory automation control room. Figure 5.5 shows the configuration of the system.

The machining stations are composed of machine centres and NC lathes. Machining processes are performed by the model FH40 horizontal spindle machining centre and the FV40 vertical spindle machining centre, both manufactured by Toyoda Machine Works.

Robots are employed in the assembly and inspection stations. The robots in the inspection stations are equipped with television cameras and perform inspection tasks by sophisticated image-processing techniques.

The operational instructions to the apparatus at each station are issued via a 5 Mbps carrier band MAP by factory automation computers

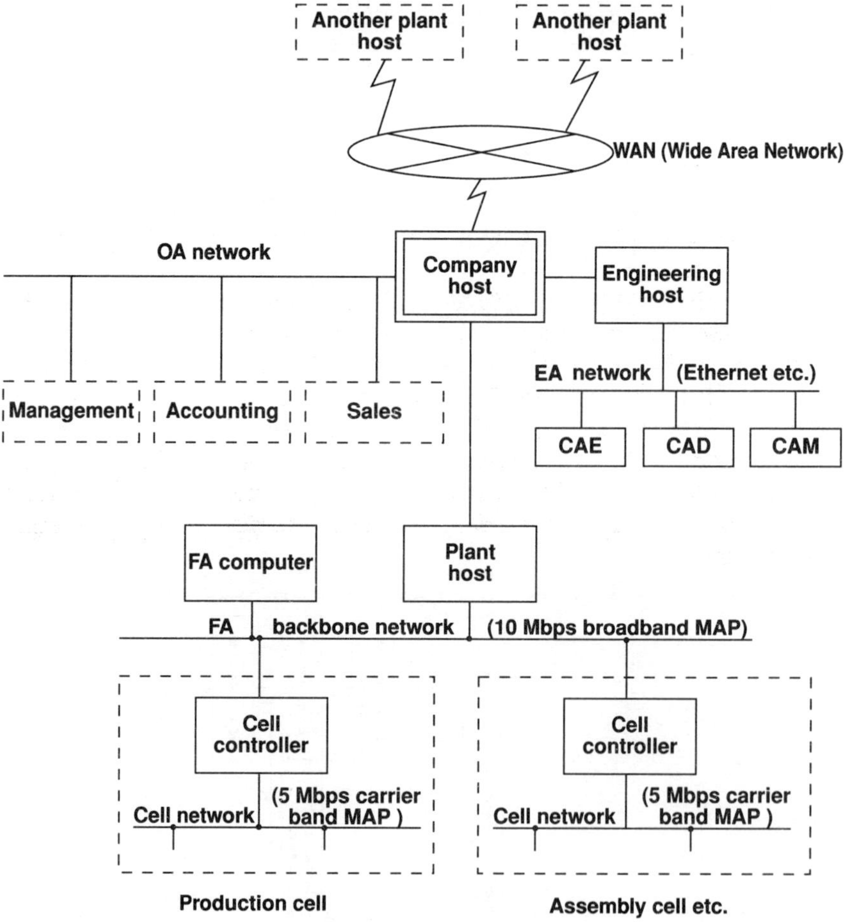


Figure 5.6 Schematic diagram of total information network.

(PCs) in the factory automation control room. These computers are also linked with CAD devices and various apparatus in other rooms. This permits practical training in a wide variety of advanced techniques such as preparation of machining data, organization of tool data, organization of measurement data, and CAD technology. The present system has thus been devised to permit acquisition of the technical expertise required for establishing factory automation systems suitable for machining in small and medium-scale industrial enterprises.

### 5.3.3 DEVELOPMENT TOWARD A CIM SYSTEM

The present system constitutes an example of the application of MAP networks to an FMS. In an overall company CIM system, however, such FMS networks would be subnetworks linked to form the comprehensive information network. This is illustrated by the schematic diagram in Figure 5.6.

## 5.4 CONCLUSIONS

The above examples have served to illustrate the practical application of MAP networks to factory automation with a view to the implementation of CIM systems.

The version 3.0 specifications for MAP were issued as an interim goal in 1988, and the general widespread introduction of MAP occured during 1991–2. As mentioned in the introduction, however, the integration of information flow through networks is an indispensable facet of the key technology required for the realization of CIM. Therefore, in combination with techniques for the creation of databases, trends in the development of MAP, which provides an international standard for LANs in factory automation, will undoubtedly draw increasingly wide attention.

# FMS for the machine industry

# 6

*Akimitsu Nagae*

Some 40 years have passed since the world's first NC machine tool was developed in the USA at MIT. Since then, numerous related technologies have been introduced with the goal of building a production system which can turn out a large variety of products with high productivity. Many such technologies have already been put to practical use.

Seeing the great potential of NC machine tools, Japanese machine-tool manufacturers have been improving their technologies, developing new functions, and making them practical from their introduction. They have succeeded in supplying users all over the world with sophisticated NC machine tools at reasonable prices. The result of this is that Japan has been the largest producer of machine tools in the world since 1982.

In 1967, Mollins of the UK announced their System 24. This system was designed to operate several machine tools under the control of a single computer. This was expected to increase productivity far beyond the level attainable by the same number of machines working independently. This turning point triggered the development of a sequence of similar production systems in the USA and Japan.

In 1981, pioneer Japanese companies such as FANUC and Yamazaki Mazak developed large-scale systems for factory-wide automation of production. These systems had the advantage of being capable of producing large varieties of parts in small-scale lots, with high productivity. These systems were eventually called flexible manufacturing systems (FMS).

*Manufacturing Automation Systems and CIM Factories*. Edited by K. Asai, S. Takashima and P.R. Edwards. Published in 1994 by Chapman & Hall, London. ISBN 0 412 48230 4

The two oil crises of 1973 and 1980 had a significant effect on Japanese industry as it is totally dependent upon imports for energy sources and raw materials. Japanese industries were sufficiently responsive to use these crises to make a significant change in the characteristics of the Japanese economy. This change involved switching from mass production and mass consumption to an individually oriented qualitative economy.

In a qualitative economy, consumers make selections according to their individual tastes. This means that products which are marketed on a uniform, mass-production basis can no longer satisfy their increasingly diverse needs. A production system which is flexible enough to quickly machine a wide variety of workpieces is needed to meet consumer preferences and bring products to the market at reasonable prices. The development of FMS is seen as an answer to this requirement.

An FMS consists of several NC machine tools which are interconnected by an appropriate materials handling system. A single computer is in overall control of the system. The following advantages are realized by FMS operation:

(a) a wide range of workpieces can be machined automatically;
(b) overall lead times are relatively short;
(c) productivity per employee is increased by unattended operation during night shifts;
(d) in-process inventory and machining costs are reduced;
(e) the system can respond flexibly to changes in products and production levels.

Japanese machine tool builders not only sell FMS, but also make extensive use of them for their own production. This chapter describes some typical examples of FMS used in the production of NC machine tools.

## 6.1 APPLICATION EXAMPLES

### 6.1.1 FMS FOR SMALL CUBIC PARTS

Small cubic parts are exclusively machined by horizontal machining centres (HMC). An HMC can be incorporated into an FMS relatively easily by equipping it with a pallet changer system. An essential requirement for machine tools which machine a wide variety of workpieces continuously, is that workpieces, tools and machining data can be easily changed. An HMC is typically equipped with these functions or they can be added to the machine as required.

Figure 6.1 shows a typical example of a small-scale FMS which

Figure 6.1 FMS for small cubic parts.

incorporates an automatic storage and retrieval system, for workpiece storage and transportation, and two HMCs. The main task of the operator, at the loading station to the right of the AS/RS, is to load workpieces onto pallets in the sequence requested by the computer. Once a pallet is loaded a button is pressed on the control panel to inform the computer that workpiece setup is complete.

A stacker crane transfers workpieces to the machine at which they are to be machined. If the machine is processing another workpiece, the pallet is carried to the stocker for temporary storage. The required tools are stored in the machining centre. If a required tool is not stored in the magazine, a tool has reached the end of its service life, or a tool is broken, new tools are sent to each machine from a central tool stocker by the tool transport system. New concepts for tool transportation and management, as shown in Figure 6.1, have been adopted in FMS. The FMS shown in the photograph is designed to allow expansion of the AS/AR and the installation of additional machines to respond to increased production requirements. The use of a vertical AS/AR has minimized the required floor space and reduced the time for changing workpieces, thus improving the operating efficiency of the system.

### 6.1.2 FMS FOR ROTATING PARTS

An FMS for rotating parts is not as flexible as one for cubic parts as it can only handle a limited range of workpieces. This disadvantage has

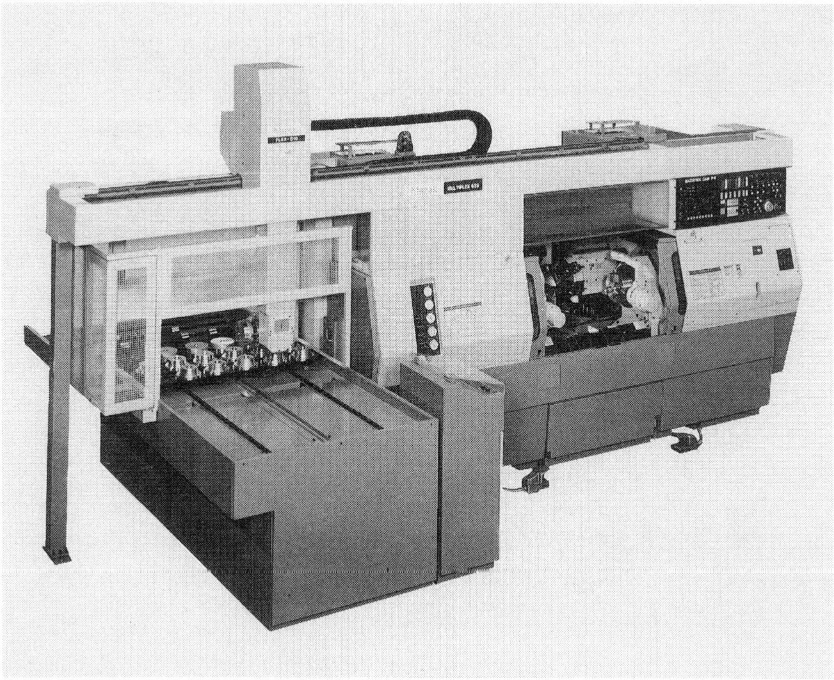

Figure 6.2 FMC for rotating parts.

evidently restricted their widespread acceptance by users. Some of the technical difficulties associated with using robots to change workpieces, and to automatically change chuck jaws, have been overcome. A standard FMS which is both flexible and economical in handling a variety of workpieces, and hence acceptable to a wide range of users, is yet to be developed. These circumstances have led to a growing tendency toward closely examining the machines currently employed in an FMS when developing new machines so as to make their incorporation easier.

Figure 6.2 shows a typical example of a flexible manufacturing cell (FMC) which incorporates a complex lathe. This can perform secondary machining operations in addition to turning. It is equipped with two opposing spindles on the same centre line, two tool turrets, and a gantry robot loader. All the required machining operations can be performed in one setup as the cell is composed of two symmetrical machining units. The internal configuration of the FMC is shown in Figure 6.3. It can be used to machine rotating parts of any configuration as it is capable of drilling and machining operations.

An increased process integration is required for performing small lot

Figure 6.3 Internal view of FMS turning machine.

production for a variety of workpieces by an FMS. It is generally accepted that this can be achieved by increasing the number of functions of individual machines. This has the effect of reducing the number of processes required. The cell described earlier has successfully achieved the goal of handling rotating parts in an FMS by combining the functionality of two machines in a single compact unit.

### 6.1.3 FMS FOR SHEET METAL PARTS

The increasing operation speed of NC machines and the progress of automation has placed a greater importance on sheet metal parts. A primary precaution for keeping machined chips and coolant contained is to provide complete protection of the machining area by means of covers. Hence sheet metal parts are becoming increasingly important components of NC machine tools. Yamazaki Mazak has installed a sheet metal FMS to increase the efficiency of sheet metal part production. This consists of CAD/CAM, laser FMS, press brake, and robot welding, painting and assembly stations. The usual batch production system has been replaced in this FMS by a flow production system. All of the sheet metal parts which are required for a single machine tool are produced at one time.

Figure 6.4 FMS for sheet metal parts.

The new production system has significantly reduced the In-process time from 90 days to five days. At the same time the amount of work in progress has been reduced. Figure 6.4 is a photograph of the laser FMS which is the heart of the sheet metal FMS. The machining performed by this FMS was previously done by punch presses. This precluded the use of a flow production system as the time loss incurred during tool setup was too large. In contrast, laser machining is ideally suited to an FMS as it requires no tool setup time.

## 6.2 CONCLUSIONS

This chapter has described the reasons why FMS is in demand in this age of a qualitative economy and has presented typical examples of FMS.

Current research projects related to production technology are concerned with computer networks. The prospects for factory automation will undoubtedly be discouraged unless due consideration is given to the fact that products are actually produced on the shop floor when designing the automation of operations. A growing importance is likely to be placed on the automation of shop-floor machines as automated machine operation is essential to producing better products with minimal labour requirements. The development of intelligent NC machine tools, and FMS peripheral equipment such as robots, is essential to translate production theory into practice.

# FMS in appliance manufacture 7

*Koji Saito*

Appliance makers, i.e. automobile and other manufacturers, are striving to improve their production technology. Their efforts focus on increasing equipment productivity and reducing labour requirements. The primary tool for achieving these goals is factory automation. Appliance manufacturers are switching to robot-based flexible manufacturing systems (FMS). FMS are therefore the major subject of production technology for a variety of industries.

The 1950s introduced the widespread mass production of appliance parts and focused on increasing the speed of moulding machines, presses and automation. When the 1960s brought about the automation of assembly processes, the focus shifted to automating and increasing the speed of the assembly and transfer of appliance parts. Although mass production and assembly-process automation differ in focus, the major theme in improving production technology was achieving mass production, i.e. pursuing high-speed technology.

The oil crisis of 1974, and the diversification of requirements, created a new social environment. Appliances were produced in yearly quantities, ranging from hundreds to thousands. The diversification of user needs led to a reduction in production quantities to the order of a few hundred units. At the same time the number of product models increased drastically. The result of this was that appliance manufacturers could no longer sustain growth as they had expanded their markets through the reduced costs and higher quality achieved via the mass production of single products. Thus appliance manufacturers were forced to take a U-turn and the main theme of their production strategy became the achievement of FMS.

*Manufacturing Automation Systems and CIM Factories*. Edited by K. Asai, S. Takashima and P.R. Edwards. Published in 1994 by Chapman & Hall, London. ISBN 0 412 48230 4

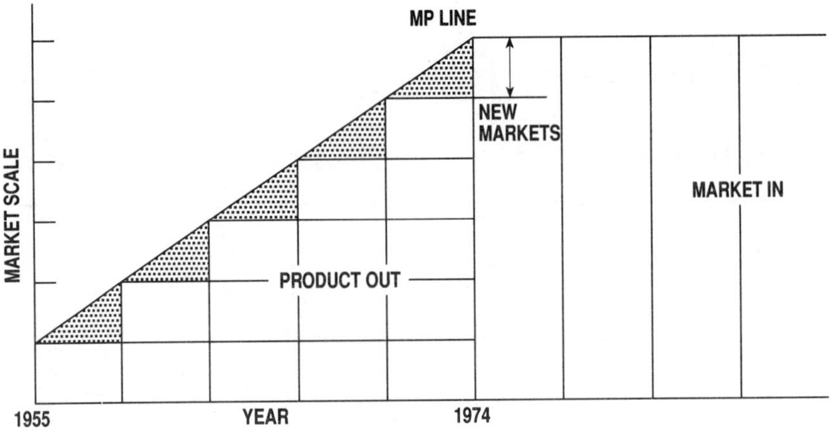


Figure 7.1 Early rapid sales growth, 1955–1974.

## 7.1 THE FMS CONCEPT IN APPLIANCE MANUFACTURE

The term FMS can be considered the antonym of mass production, i.e. high product mix, low volume production. In the Matsushita Group, however, it was considered necessary to add the additional factor of early start-up. Figure 7.1 shows that during the period of rapid growth prior to 1974, Matsushita were able to increase sales through market expansion. This was in spite of the fact that they were one of the later companies to set up a production system. In the low growth period following the oil crisis, new markets were scarce and late-starting companies could not achieve the same success.

FMS allows:

(a) high product mix, low volume production (i.e. easy change over between models);
(b) rapid production system start-up (as well as rapid modification of production systems).

## 7.2 FMS EVALUATION MEASURES

The conventional evaluation measures for the effectiveness of production equipment investment are inadequate for the evaluation of FMS. Therefore new evaluation criteria must be determined.

The experience of the Matsushita group has shown that good FMS are adaptable to changing production conditions, for example changing quantities, models and processes. This is best achieved by using equipment with low modification costs. These low modification costs are themselves achieved through high equipment productivity in production, and low modification times. FMS can be evaluated by the

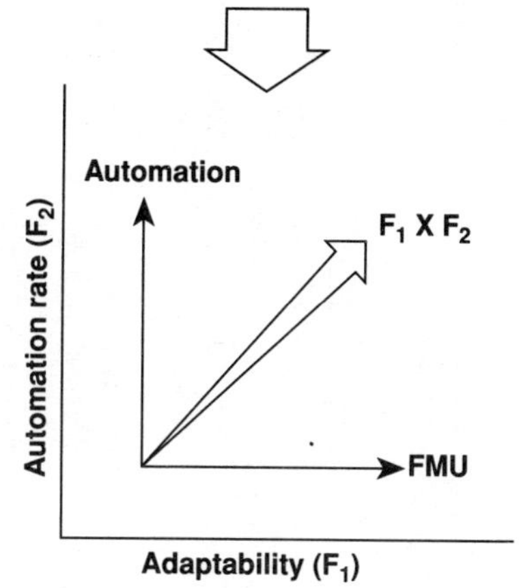


Figure 7.2 Production equipment FMS evaluation measures (see Table 7.1).

measure which is the product of an investment effectiveness index, an adaptability index, and an area utilization index (Figure 7.2, Table 7.1). These factors should be evaluated and examined prior to investment in equipment.

In summary, good FMS can be characterized by:

(a) advantageous equipment investment;
(b) low modification cost during changes in production conditions;
(c) compact equipment.

## 7.3 PROGRESS TOWARDS FMS REALIZATION

A correct approach is essential for achieving good FMS. When a production system is designed to suit conditions dictated by the production site (such as product models, product change over conditions, time periods for changing to new products) the result is increased equipment costs and poor effectiveness of investment. It is a better approach to modify the production conditions prior to the determination of FMS equipment specifications.

### 7.3.1 PRIOR ADJUSTMENT OF PRODUCTION CONDITIONS

The principle of 'simple is best' is applied to the adjustment of production conditions as follows:

(a) Productivity checks for products which have several models:
   (i) standardize all parts;

**Table 7.1** Production equipment FMS evaluation measures

| Measure | $I \times F \times A$ | | |
|---|---|---|---|
| | Investment effectiveness index | Adaptability index | Area utilization index |
| Index | $\frac{\text{Effective cost}}{\text{Investment cost}}$ | $E_1 \times F_2$ | $\frac{\text{Ideal minimum area}}{\text{Area occupied by equipment}}$ |
| Calculation conditions | Effective cost<br>Labor cost<br>Material cost<br>Quality improvement cost<br>Investment cost<br>Equipment cost<br>Operating cost (power maintenance) | $\text{Adaptability} = \frac{\text{FMU cost}}{\text{Investment cost}}$<br>$\text{Automation rate } (F_2) = \frac{\text{Automated processes}}{\text{Standard processes}}$ | Ideal minimum area<br>(Projected area of product × Number of processes)<br>Area occupied by equipment<br>Projected area of equipment<br>Maintenance space<br>Part supply space |

   (ii) if complete part standardization is impossible, give all parts common dimensions to the maximum possible extent;
   (iii) if parts cannot be adequately standardized, at least ensure uniformity in part manufacture and installation.
(b) Organization of the production control system:
   (i) part supply methods and timing;
   (ii) production data consolidation and collection methods.
(c) Organization of the quality control system.

Once common specifications for inspection methods and equipment have been achieved, it is necessary to determine the specifications for the FMS production equipment. Since two- to three-year forecasts are necessary at this stage, this cannot be the sole responsibilty of the production engineering department. This process should also involve the planning department, the product development department and management.

### 7.3.2 ACTUAL FMS COMPOSITION

The actual FMS equipment composition should be determined after production condition adjustment and the determination of basic specifications. Actuator systems are normally converted to robots so as to provide greater equipment flexibility. The trick with FMS, however, is to provide flexibility to systems which do not employ actuators. This objective is accomplished by the use of modular equipment, such as that shown in Figure 7.3. Thus a system is formed from a combination of individual unit modules, element modules and stand-alone units.

## 7.4 POSITIONING AND APPLICATION IN FMS

Figure 7.3 shows that robots are the unit modules of FMS. It is therefore essential to select robots which have manageable equipment investment costs. Cartesian robots are ideal in that they are low cost and have outstanding operability and safety characteristics. Current robots can be simply regarded as a type of actuator, and hence should be selected according to their intended usage.

## 7.5 CONCLUSIONS

As robots are simply one form of actuator, it is important, when working towards FMS, to gain an understanding of the capabilities of production systems based upon them. This understanding should be used to forecast the production conditions two to three years into the future. This should include the greater sensor capabilities which will

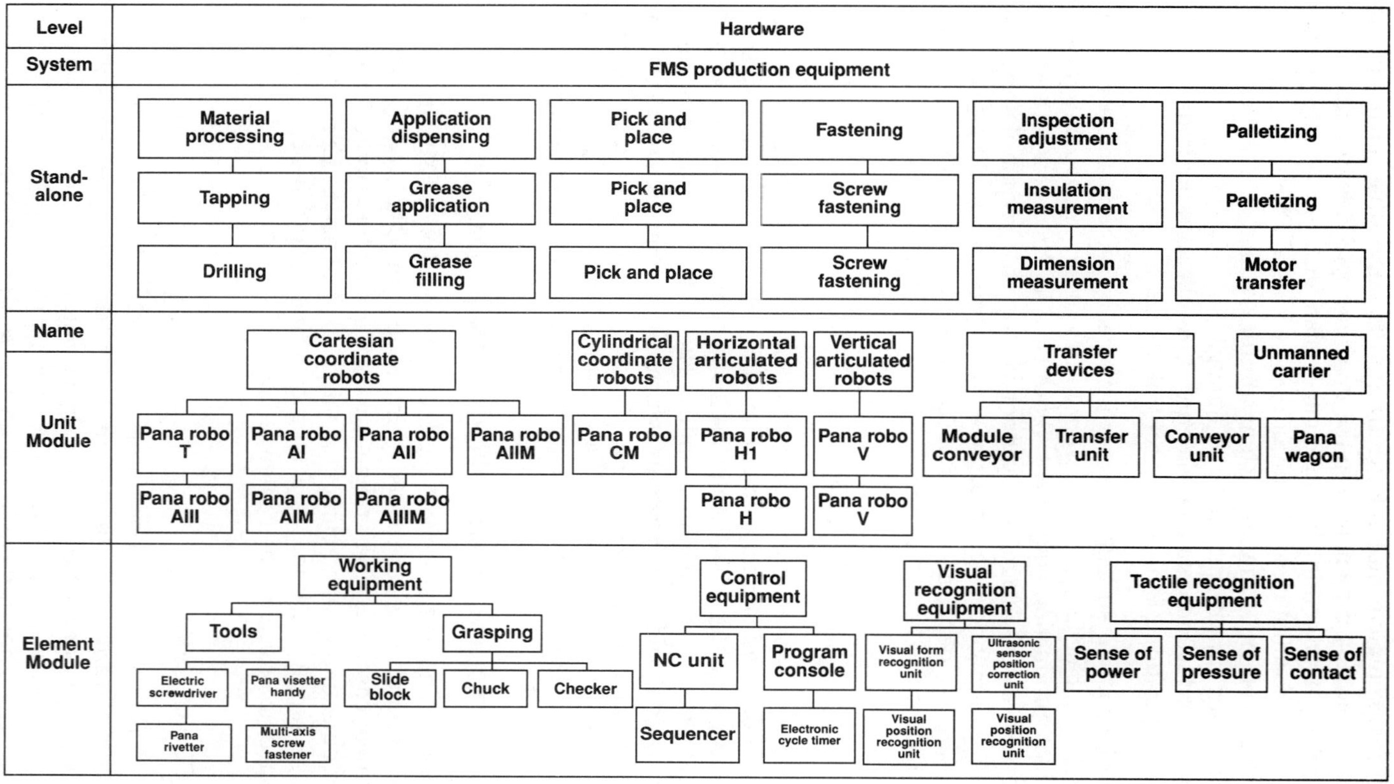


Figure 7.3 Modular approach to system configuration.

be developed for future robots. It is anticipated that greater sensor capability will shift robots from the unit module level to the stand-alone level (Figure 7.3). It is uncertain what the corresponding effect will be at the FMS level. The equipment manufacturer Matsushita views this effect as the necessary expansion of the unit range and element modules (Figure 7.3). They are working in the direction of upgrading the fundamental technology to meet this.

# Factory automation for control motor manufacture

# 8

*Toshio Kojima*

The significant achievements which can be obtained through factory automation are examined in this chapter by looking at the Fanuc control motor factory. Industrial robots have been introduced into all processes of this factory, from the machining of parts through to assembly, in order to achieve higher efficiency in production.

The Fanuc factory has a full-scale factory automation system which has integrated machining and assembly cells along with an automatic warehouse and unmanned carriers. Each cell includes one or two robots, giving a total of 115 industrial robots in operation.

This system was created to fully integrate all of the processes involved in the manufacture of control motors. This includes the receiving of parts and materials, machining of parts, assembly, final inspection, and packaging.

The conventional automation of factories using factory automation and flexible manufacturing systems has involved only about a dozen machines. These are usually engaged in a limited range of processes, or in the processing of a limited range of parts. The Fanuc factory automation, however, is on a factory-wide scale. This has been achieved by developing a large range of automation technologies relating to industrial robots. In particular the automation of motor assembly requires that robots be used to their full potential.

The Fanuc factory, in keeping with its original objectives of cost and production efficiency, has only 60 employees. This figure includes managers and production staff. Its monthly production is in excess of 10 000 units of 40 models of servo-motor and spindle motor (Figure 8.1).

*Manufacturing Automation Systems and CIM Factories.* Edited by K. Asai, S. Takashima and P.R. Edwards. Published in 1994 by Chapman & Hall, London. ISBN 0 412 48230 4

Figure 8.1 Examples of the control motors manufactured by the FA system at the Fanuc control motor factory.

## 8.1 FACTORY LAYOUT

Figure 8.2 shows the overall layout of the Fanuc factory. Parts are machined on the first floor, and assembled on the second floor. An automatic warehouse provides a link between the first and second floors. Automation is achieved through the use of a wide variety of CNC machines and robots. This has resulted in an outstanding level of flexibility.

The materials transfer both inter cell and between cells and the automatic warehouse is completely automated using unmanned carriers. A central control room, on the second floor, contains the controlling computer system and central monitoring systems. This is responsible for the constant monitoring of the operating conditions throughout the factory.

The following measures have been taken to minimize the personnel required:

(a) Sophisticated and versatile industrial robots have been introduced into all areas.

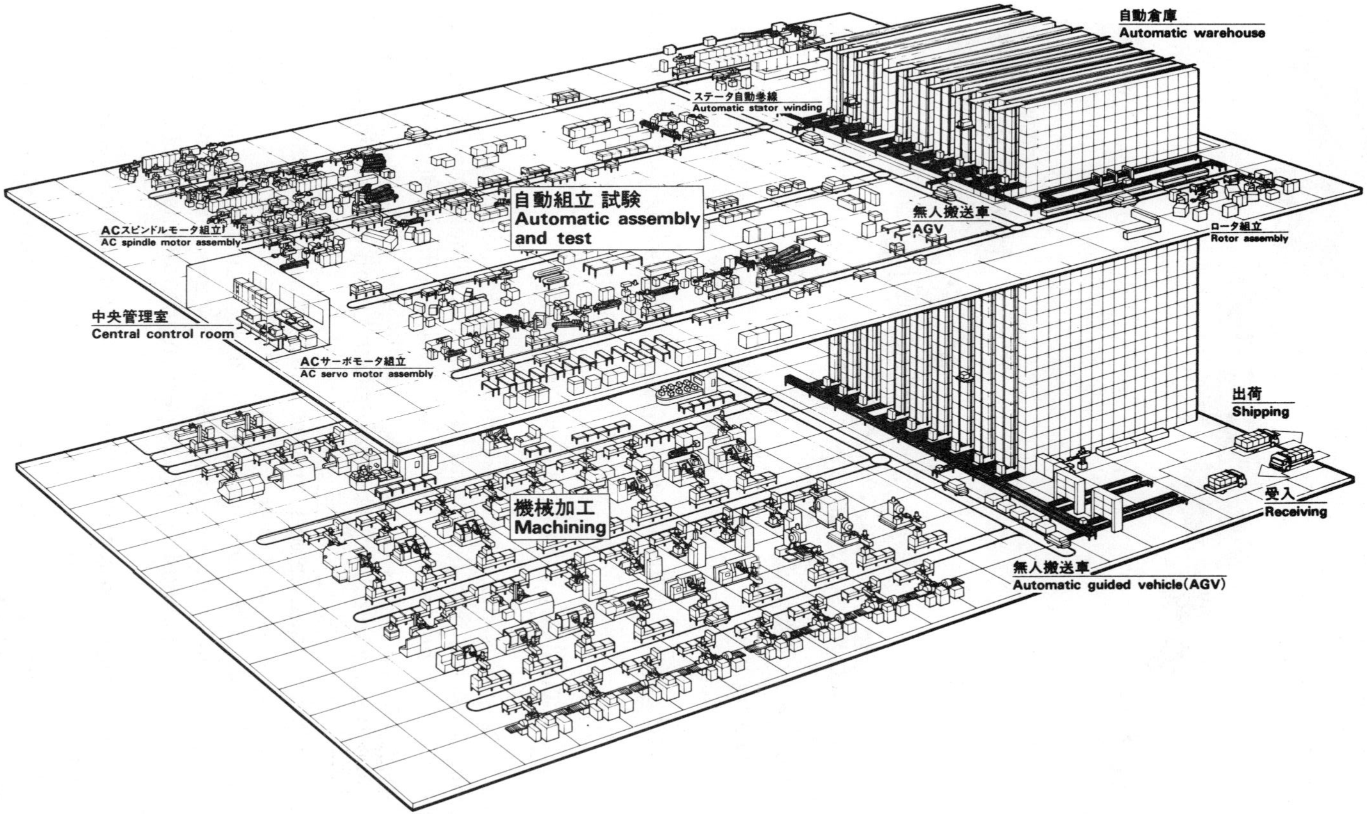


Figure 8.2 Layout of Fanuc control motor factory.

(b) Parts pallets have been standardized by type. This enables robots to load and unload parts directly to or from the pallets.
(c) All of the factory's processes have been automated, from the initial reception of parts through machining and assembly to final inspection and shipment. This results in a 24-hour operation which is unmanned at night. Unmanned daytime assembly is to be introduced soon.

Figure 8.3 shows the flow of parts within the factory. Upon arrival, parts and materials are individually placed onto standardized pallets, which are taken to the automatic warehouse. Parts which require inspection on arrival are then sent to the parts inspection area. They are not delivered to the machining lines until inspection is complete. Parts which do not require inspection, and those which do and have passed inspection, are stored in the automatic warehouse until they are required on the machining line. This is controlled by the overall processing program.

When the machining schedule requires certain parts for a particular machining cell, the appropriate parts are automatically extracted from the automatic warehouse and delivered to the designated cell. Once the correct number of machining operations have been performed on the designated number of parts as directed by the schedule, the cell proceeds to process its next assignment. The pallet of machined parts is returned to the automatic warehouse pending its next process. Following the completion of all machining processes, the finished parts are inspected. Parts which pass the inspection are ready for assembly on the assembly floor.

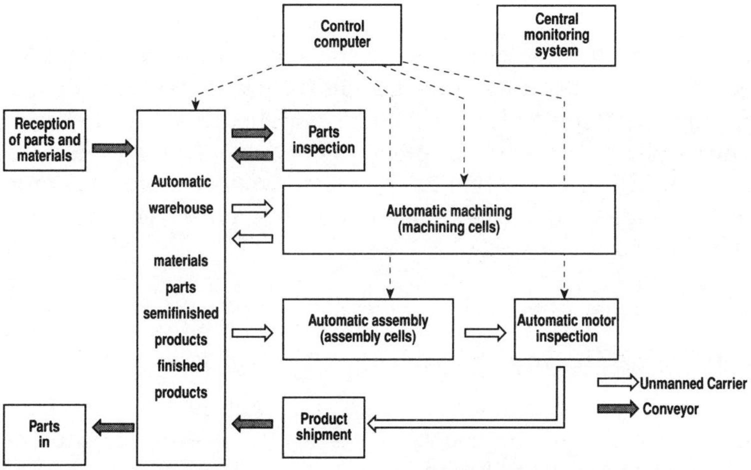


**Figure 8.3** Factory flowchart.

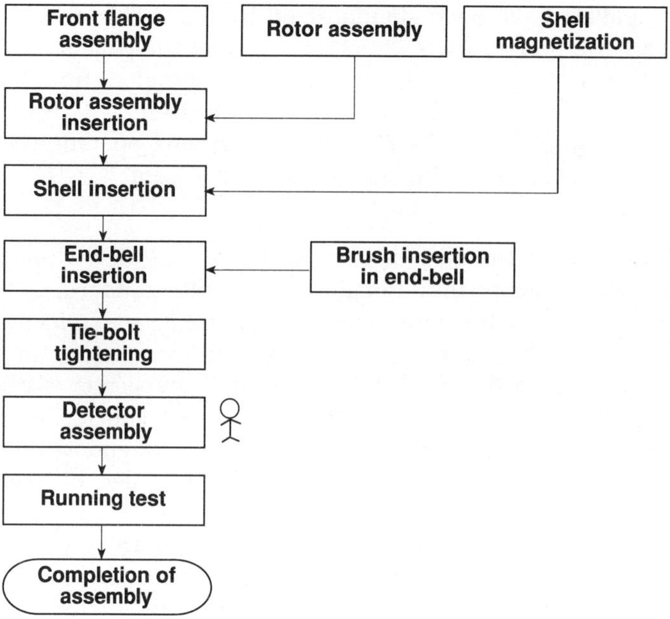


Figure 8.4 D.C. servo-motor assembly procedure.

For the assembly operation, all of the parts which are required for the assembly of specific motor types are extracted from the automatic warehouse and delivered to the appropriate assembly cells. Parts are picked up directly from the pallets by an assembly-line robot, and automatic assembly proceeds. Figure 8.4 shows an example of the assembly procedures for d.c. servo-motors. Computer-controlled automatic inspection is performed after the assembly operation is complete. Products which pass inspection are automatically packed, in the packing area, and transferred to the automatic warehouse to await shipment.

The automatic warehouse contains all of the items required for factory operation. This includes materials, parts, units and finished products. It also contains tools, jigs and maintenance parts. The warehouse is fully computer-controlled.

## 8.2 AUTOMATIC MACHINING

The parts machining floor, as shown in Figure 8.5, consists of 60 machining cells and 55 robots. Parts are transferred to the cells by means of two unmanned carriers.

The machining cell is the smallest unit capable of continuous, auto-

Figure 8.5 Parts machining plant.

matic machining for small-lot, wide-variety production. The minimal configuration of a machining cell consists of one NC machine tool, one robot, and a pallet station.

Machining data on parts specification and quantities is supplied to each cell by the central computer system. The specified parts are transported to the pallet stations by unmanned carrier. Each pallet station has three pallet platforms: one for parts being machined, one for completed parts, and one for the parts awaiting processing.

Each part is individually removed from its pallet and set directly on the NC machine tool by robot. The robot replaces the part back on its pallet after machining. Once all of the parts of a pallet have been machined, the pallet is transferred to the finished parts platform, ready to be returned to the automatic warehouse by unmanned carrier.

Cutting dust generated during the machining operations is automatically cleaned away. The system automatically compensates for aberrations in tool dimensions due to wear. This allows continuous unmanned operation for many hours at a time.

The processing of machining chips is also simple. They are collected in a standardized bucket. When a bucket is full it is placed on a pallet station where it is taken to the chip stocker by unmanned carrier. An empty bucket is carried to the pallet station as a replacement.

Machining operations on large parts which are too heavy for robot handling are also automatically processed. They are prepared in advance on pallet pools which can accommodate between six and ten pallets each. Automatic machine centring and position compensation allow automatic machining without manual intervention.

## 8.3 AUTOMATIC ASSEMBLY

The motor assembly plant contains 35 assembly cells and 60 robots. These cells perform the automatic assembly of all motor types with the exception of some highly specialized models. There are four assembly lines, each capable of the automatic assembly of a wide range of motor models. In addition, the assembly lines are devised to require almost zero setup procedures, even when different models are assembled on the same line. This is made possible by the versatility of the assembly system.

Each assembly line has four or five assembly cells which perform all of the operations from robot machining to the automatic inspection of finished models. Each line uses between seven and ten robots. These robots are interconnected through their respective controllers so as to achieve integrated operation throughout the line. Parts are automatically supplied to the assembly lines in a similar way to that of the machining lines.

Figure 8.6 Assembly plant.

Conventional assembly plant operators are responsible for checking for defective parts and malfunctions in the assembly process. Automatic error detection is necessary in automatic assembly operations as there are, by definition, no operators.

Highly skilled operators find that large numbers of assembly procedures are cumbersome. Robots with video sensors, however, can perform these tasks automatically. Tasks which require three-dimensional assessment of workpieces require the use of two video cameras for automatic assembly.

Figure 8.6 shows a partial view of the assembly plant.

## 8.4 AUTOMATIC INSPECTION

After assembly, robots set the finished motors on inspection platforms. They then undergo computer-controlled inspection. Inspection data is stored permanently on disc.

Motors which fail inspection are set aside by a robot. Only motors which pass inspection are placed on pallets and sent to the automatic warehouse.

## 8.5 MAJOR TECHNOLOGICAL DEVELOPMENTS

A very diverse range of new technologies has been developed for the Fanuc factory. The most important of these are discussed in the following subsections.

### 8.5.1 THE DEVELOPMENT OF AUTOMATION TECHNOLOGIES

The aim of the control motor factory is to achieve totally unmanned operation. In pursuit of this goal, numerous technologies have been developed. Robot technology is at the core of automation technology. In particular, a vital factor for the effective use of robots is the development of automatic robot peripheral technologies such as end-effectors.

Approximately 40 different robot hands (end-effectors) were developed to cater for the diverse specifications of parts profiles and factory machines. Of particular importance was the development of a specialized hand capable of directly removing parts which are stacked on a pallet. Hands capable of tightening screws and applying sealants were developed for assembly applications.

A wide variety of devices were developed to maintain a high precision in machining. These include:

(a) devices which clean the centre holes of machines and apply oil;
(b) devices which clean away cutting dust after machining;

(c) devices which automatically remove burrs after machining;
(d) devices which measure the dimensions of machining parts (for example, tools) and carry out dimensional compensation.

Robots equipped with video cameras were developed to accomplish complex assembly tasks. These have already proved very successful in practical applications.

It is essential to ensure that assembly tasks have been completed accurately and according to specification on automatic assembly lines. A wide range of sensors has been developed to meet this requirement.

Many different types of motor are processed on each automatic assembly line. To overcome the problems normally associated with this scenario, a system was designed which requires minimal setup, even with different motor types.

Finally, various automatic inspection systems were developed and installed to automate all of the inspection processes with the exception of the inspection of the motor exterior.

### 8.5.2 FACTORY AUTOMATION SYSTEM DESIGN

The Fanuc control motor factory was conceived as a factory automation application which can in principle operate without staff. Many measures have been taken to achieve this goal. Some of the unique systems which were developed for the machining and assembly processes are as follows:

(a) a system that automatically supplies the exact quantity of parts as required by the pre-planned machining schedule;
(b) a system that automatically supplies all of the parts required for assembly given a motor type and production quantity;
(c) a system that determines which parts require inspection and automatically transfers them to inspection stations;
(d) a system that transports a full bucket of machining chips to a collecting station and returns an empty bucket.

In addition, a comprehensive operations control system is used. This has a high-level host computer which receives and issues data on daily scheduling, production records, and the flow of goods into and out of the warehouse. A automatic system is also used to process all orders and to receive and inspect procured parts and materials. Regular warehouse inventory checking is performed automatically and rapidly. Finally, a system was developed which performs central monitoring of all factory operating conditions. This allows unmanned factory operation.

### 8.5.3 ASSEMBLY ROBOTS WITH VIDEO CAMERAS

Robots equipped with video cameras were developed to maximize the operating efficiency of the motor assembly plant.

#### (a) Application to stator-wire lead extraction

The spindle motor assembly line of the control motor factory includes a process during which the stator-wire lead is extracted. This forms part of the process which attaches a bracket to the motor after the rotor has been attached to the stator. The ends of several stator wires are tied together, covered with a vinyl cap, and pulled away at an angle from the motor. The bracket, which is held by a magnetic hand attached to the robot wrist, is then attached to the stator. While this is being done, the stator wire must be passed through a hole in the side of the bracket. This process is referred to as 'wire lead extraction' and is shown in Figure 8.7.

In order to extract the wire using only the robot, the end of the wire must always be confined to a fixed range of positions. It is difficult to maintain this positional constraint owing to the flexibility of the wire and the tightness of the range of positions. This is true even if a preforming operation is carried out prior to assembly. In addition, the jig used in the previous process, to insert the rotor into the stator, presses down on the wire. This results in each motor's stator-wire being in a different position.

Visual sensors are employed to detect the position of the end of the wire and to correct the robot's motion. Due to the fact that the wire

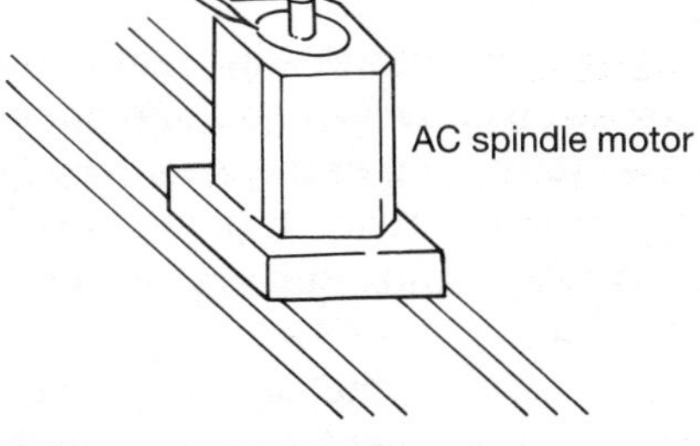


**Figure 8.7** Wire lead extraction.

Figure 8.8 Automation of wire lead extraction by robot equipped with video sensors.

end position varies in three dimensions, and the fact that the length of the wire is not constant, the visual sensors must be able to locate the wire end with a high degree of spatial accuracy. To achieve this, video cameras are installed both above and to the side of the assembly station. In addition to eliminating the need for manual shaping of the wires, the introduction of visual sensors has resulted in nearly perfect wire extraction using robots.

This application is shown in Figure 8.8.

**(b) Application to tie-bolt nut tightening**

During the servo-motor assembly process, the tie-bolts are tightened prior to the detector being attached, as shown in Figure 8.4. This process starts with the attachment of the rotor, shell and end bell to the front flange. The tie-bolts are inserted into the top of the end bell as far as the front flange. Washers and nuts are put onto the tie-bolts and then tightened, as shown in Figure 8.9.

The end bell has long holes for the tie-bolts. These holes are used for adjusting the motor in a subsequent process. The tie-bolts, which are long steel rods, are threaded at both ends. The tie-bolts are initially

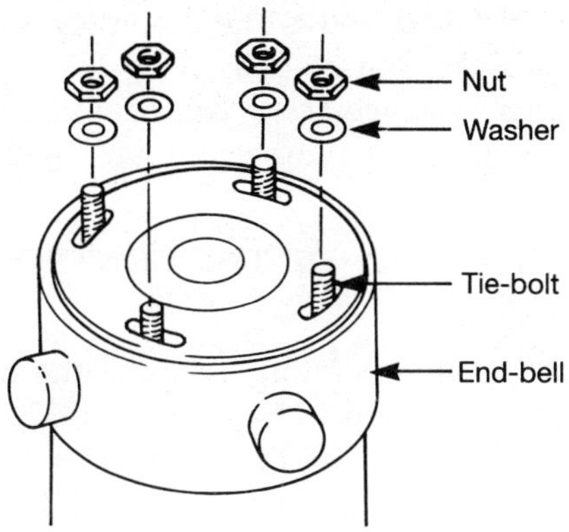


Figure 8.9 Tightening of tie-bolt nuts in servo-motor.

inserted into the holes and screwed tightly into the front flange. After this operation the positions of the tops of the tie-bolts can be variable. Even though the robot hand which tightens the nuts has been specially designed to give the required accuracy, inaccurate tightening can still occur. Visual sensors are provided to prevent this occurrence.

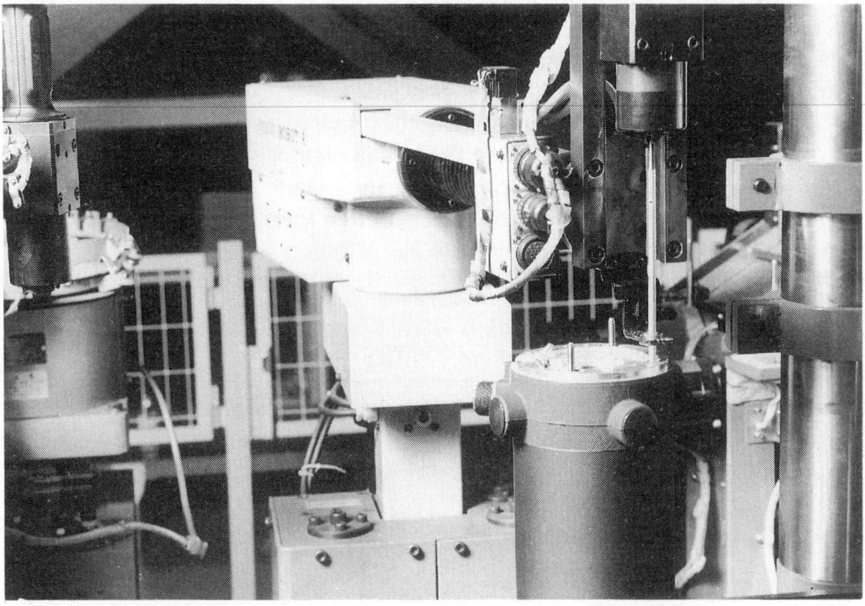

Figure 8.10 Automation of tie-bolt nut tightening by robot equipped with video sensor.

A video camera, installed above the assembly station, detects the positions of the tops of the four tie-bolts. This data is transmitted to the robot which performs any necessary compensation. This technology has eliminated all tie-bolt tightening errors (see Figure 8.10).

## 8.6 ECONOMIC EFFICIENCY OF THE CONTROL MOTOR FACTORY

A comparison between the operating status of the new control motor factory, which began operation in September 1982, and the previous conventional factory is shown in Table 8.1. The inauguration of the new factory enabled a reduction in personnel from 82 to 39. The number of robots, in contrast, increased from 32 to 115. Production volume also increased from 6000 to 10 000 motors per month. The result of this was that the productivity per worker jumped from 73.1 motors to 256 motors per month, representing a 3.5 fold increase.

## 8.7 CONCLUSIONS

The Fanuc control motor factory has been equipped with a fully integrated factory automation system. This has been developed to automate all manufacturing processes from material machining through assembly to inspection and shipment.

Previous automation systems have been applied to isolated operations such as machining. In contrast, this system was the first of its kind to achieve automation in all processes.

A highly adaptable system which combines very flexible robots with numerous lathes and machining centres is ideally suited to the machining and assembly of parts produced in a wide range and in small quantities. Thus this system is much better than systems which only automate isolated machining centres or lathes.

**Table 8.1** Comparison of old and new factories. Productivity increase = 256/73.1 = 3.5

| | Old factory | New factory |
|---|---|---|
| Number of workers | 82 | 39 |
| Number of robots | 32 | 115 |
| Production volume (units/month) | 6000 | 10 000 |
| Production value (¥bn/month) | 1 | 16 |
| Productivity (units/person-month) | 73.1 | 256 |

The factory is also designed to handle automatically certain tasks which were formerly dependent upon manual labour. Examples of this are the removal of machining chips and washing.

The Fanuc control motor factory's factory automation system incorporates many new technologies and is likely to make a major contribution to future progress in the automation of the machine industry.

# CAD/CAM in the automobile industry 9

*Toshiaki Takagi*

Automobile manufacturers are accelerating the development and use of computer-based tools to improve product quality and shorten development times. The computer-based tools include computer-aided design and manufacture (CAD/CAM), computer-aided engineering (CAE), and computer-aided testing (CAT). The architectures and philosophies behind such systems are very application dependent. Mazda's CAD/CAM system, which has been under development and in use since 1979, provides integrated hardware and software to support planning, design, engineering, and production preparation activities. This has evolved from the conventional two-dimensional designs, based on hard copy drawings, to three-dimensional conceptualization.

This chapter gives an overview of Mazda's CAD/CAM system, its applications, and the impact of its implementation.

## 9.1 PRODUCT DEVELOPMENT SYSTEMS

Figure 9.1 shows the product development processes and the various support systems which are involved in the development of a vehicle. The development starts with conceptualization, followed by a product planning stage. A basic vehicle layout is formulated and divided into blocks for the major features such as the engine compartment, boot, and passenger compartment. In addition to this layout work, sketches and a clay model are produced from which body drawings are prepared. This information is used to generate engineering drawings which are used in turn to build a prototype. The prototype is tested and the test results fed back to the design departments. The engineering drawings are updated to accommodate any changes borne out

*Manufacturing Automation Systems and CIM Factories*. Edited by K. Asai, S. Takashima and P.R. Edwards. Published in 1994 by Chapman & Hall, London. ISBN 0 412 48230 4

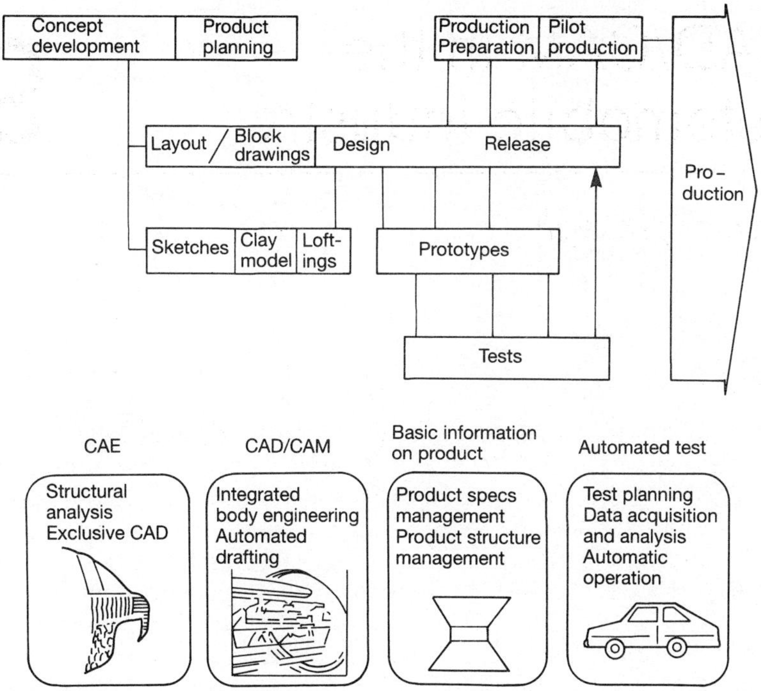


**Figure 9.1** Product development system.

of the prototype testing. The finished prototypes are released to production preparation so that the appropriate jigs and fixtures can be manufactured. A pilot production phase is performed to eliminate any remaining problems prior to the vehicle going into full production.

The computer-based systems which are currently employed in these processes are: CAD/CAM; prototype management; basic product information; and computer-aided testing. All of these systems improve design productivity, upgrade design and product quality, and shorten development times.

## 9.2 DEVELOPMENT OF CAD/CAM SYSTEMS

### 9.2.1 BACKGROUND

Mazda began research and development into CAD/CAM in 1966. They employed a total engineering concept from the very outset and in early

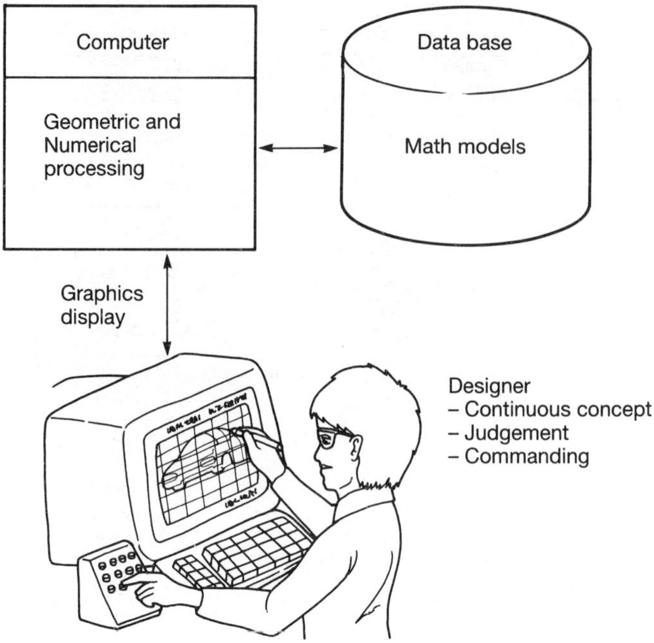


Figure 9.2 CAD/CAM system: a tool for the designer.

1970 developed a batch-type processing system (GNC1). This assisted in many operations, from the three-dimensional measurement of clay models to the manufacture of stamping dies. GNC1 was developed into an integrated CAD/CAM system (GNC2) in the period 1978 to 1983. This involved upgrading the functionality to support both basic and detailed design, and for NC processing.

### 9.2.2 PHILOSOPHY BEHIND MAZDA'S CAD/CAM

Mazda's CAD/CAM system was not intended to automate the design process, rather it was meant to be a tool to assist design engineers. In particular it must enable a designer to interact with a graphical display as shown in Figure 9.2. GNC2, based on GNC1, incorporates the latest hardware and software technologies. The CAD and CAM aspects of the system were integrated by providing common database and database access software as shown in Figure 9.3.

### 9.2.3 BASIC SHAPES AND FUNCTIONS

Figure 9.4 illustrates the basic shapes used by GNC2. Table 9.1 lists its basic functions and applications.

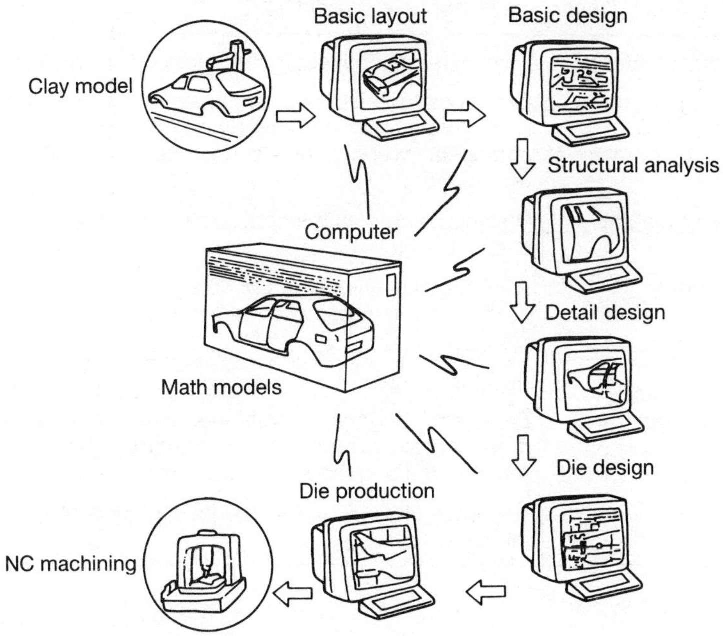


**Figure 9.3** Integrated processing system.

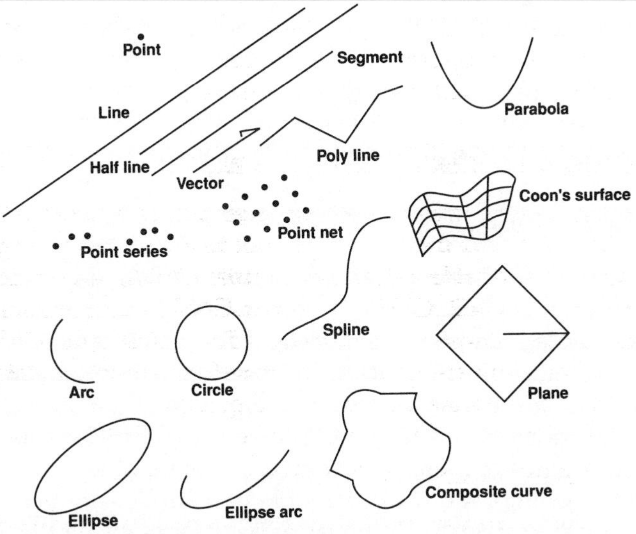


**Figure 9.4** Basic shapes used in GNC2.

**Table 9.1** Basic functions and applications of GNC2

| | |
|---|---|
| Definition of features | Point, straight line, circle (arc), free curve, surface etc. |
| Processing of features | Cross point/intersection, offset, surface, projection to plane, separation/join, rounding/trimming, composition, copy/move, surface development/ winding, etc. |
| Analysis of features | Distance, angle, line length, area, volume, properties of section, etc. |
| Applications | Kinematics simulation, FEM data preparation, visual field, manikin, suspension geometry, etc. |
| NC functions | 3D profile cutting, 3D arc cutting, interference avoidance, etc. |
| Support system functions | Enlarging, reducing, displacing, revolving, database management, screen control, temporary clear/ retrieve, drafting machine output, etc. |
| Others | Database utility, CADAM interface, other system interface |

## 9.3 IMPLEMENTATION OF CAD/CAM

### 9.3.1 OPERATING EQUIPMENT

The GNC2 operating equipment is shown in Figure 9.5. The host computers are dedicated to CAD/CAM. The IBM 5080 terminals are used as the graphic display medium in all departments. Numeric data input is provided by three-dimensional measuring devices and digitizers. Automatic drafting machines and plotters are used for data output. Users can choose between on-line and off-line devices according to their accuracy, size and turn-around time requirements.

### 9.3.2 APPLICATIONS

Figure 9.6 shows the applications of the CAD/CAM system in the planning stage. Mannequins are provided as a design aid so that head clearance and leg room can be investigated. This information is used in the basic layout guidelines used by the design department. Another CAD/CAM application is the generation of a line drawing of the body's outer panels from the measurements taken from the clay model. This is shown in Figure 9.7. This line drawing defines the principal inner-panel line drawings and the inner and outer surface profiles. Figure 9.8 shows how the rearward visibility, as seen through the fender mirror and the wiped windscreen area, can be determined. The raising and

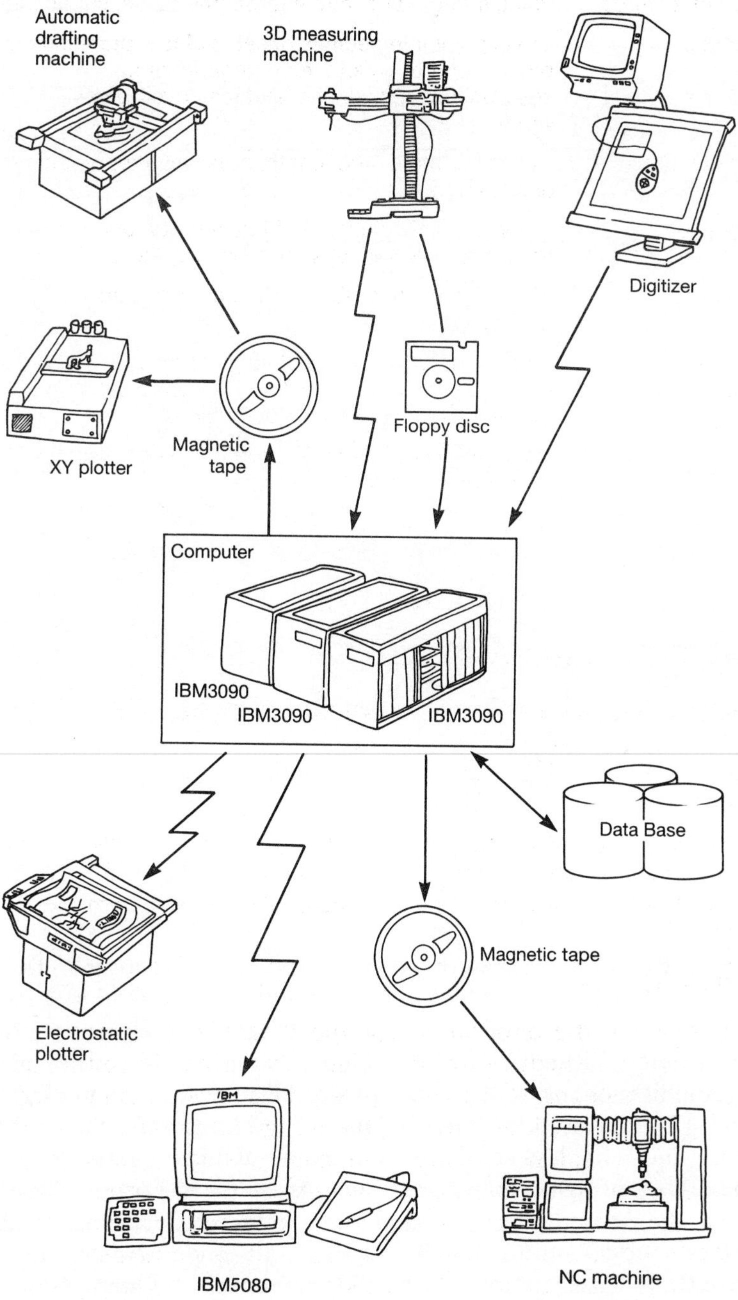


Figure 9.5 Configuration of GNC2 equipment.

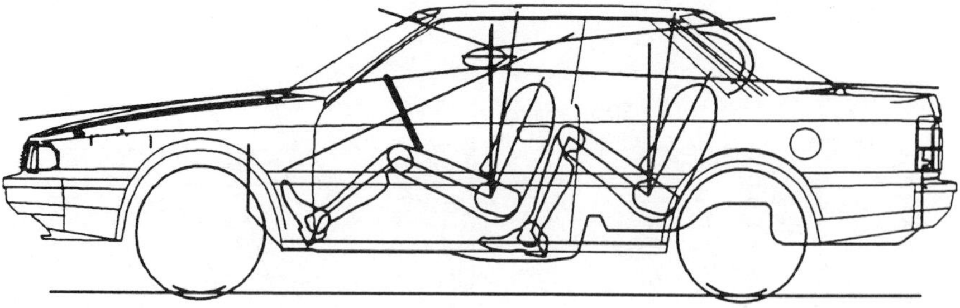

Figure 9.6 Diagram used for review of passenger space.

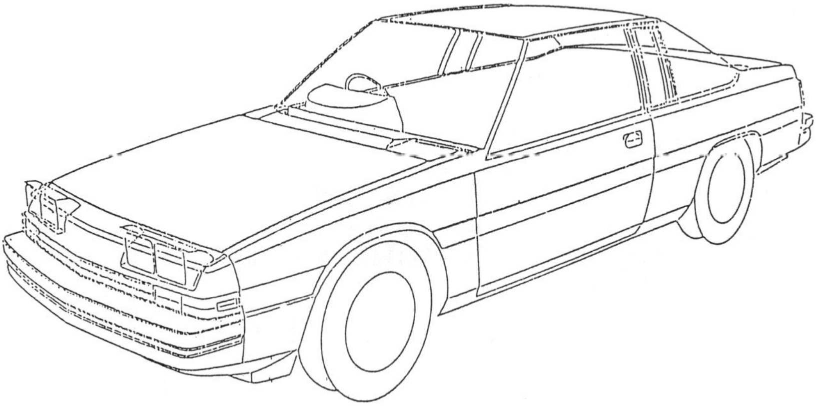

Figure 9.7 Outer panel line drawing.

lowering paths of the rear door glass are shown in Figure 9.9. These paths are used to lay out a guide so as to avoid body-glass interference. Figure 9.10 shows the input mesh data for structural analysis which is generated automatically from the body surface data. This data, after partial modification, is sent to the structural analysis system. Figure 9.11 shows the surface of a stamping die for the front fender path. The use of graphical displays, along with a greater ability to define surface area, has made it possible to digitize the excess metal portion, the die face, and the shaping face. This permits direct NC cutting of the die surface. The NC cutting paths for the front fender die are shown in Figure 9.12. These paths can be determined even when 20 or 30 surfaces are involved.

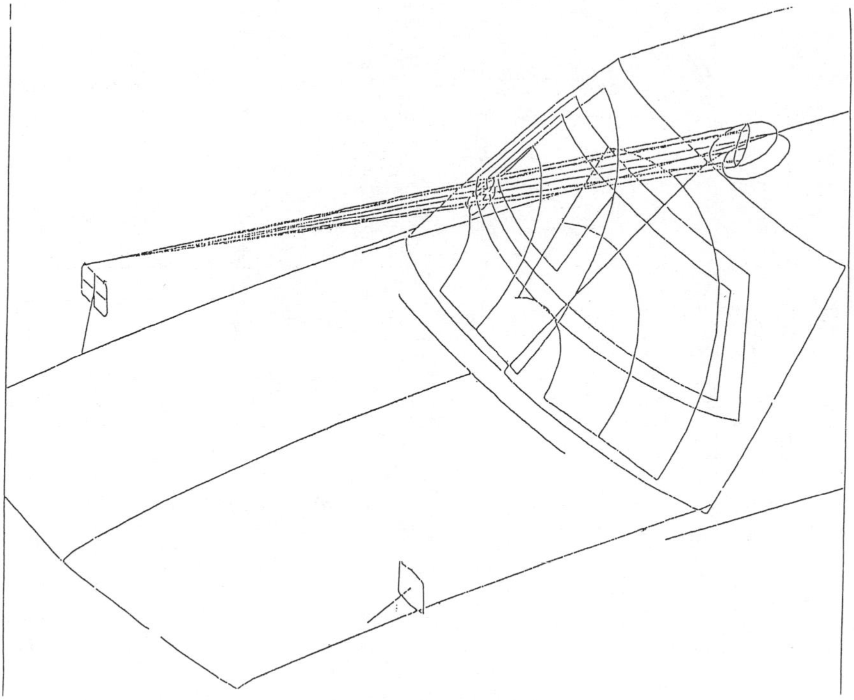

Figure 9.8 Line drawing for reviewing rearward visibility and wiped area.

9.3.3 BENEFITS

The early stages of Mazda's transition to CAD were not without problems. One such problem was the rejection of the idea by some of the designers and draughtsmen. Some of these personnel were accustomed to two-dimensional thinking and design. They were not enthusiastic about adapting to use the three-dimensional conceptualization concepts of GNC2 (Figure 9.13). A GNC2 promotion office was set up which had a consultant in each area. The new technology was introduced and demonstrated along with a system to determine the users' operating levels. Designers became more appreciative of the system as they discovered applications for the design tool. In fact some designers stated that they no longer wished to use conventional drafting methods. Designers are now demanding that the system's functionality be strengthened and that the number of graphical displays be increased.

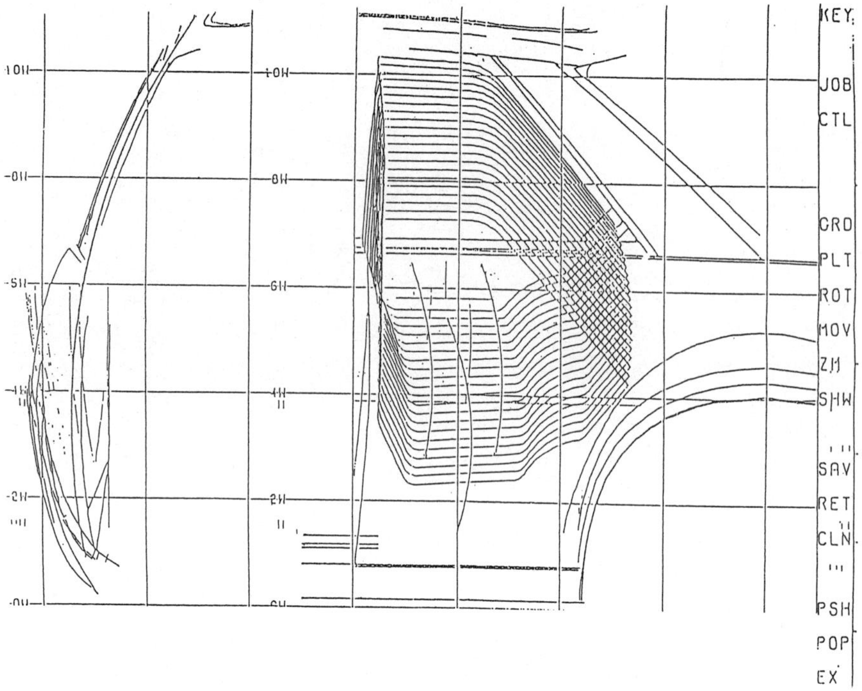


**Figure 9.9** Reviewing raise/lower path of rear door glass.

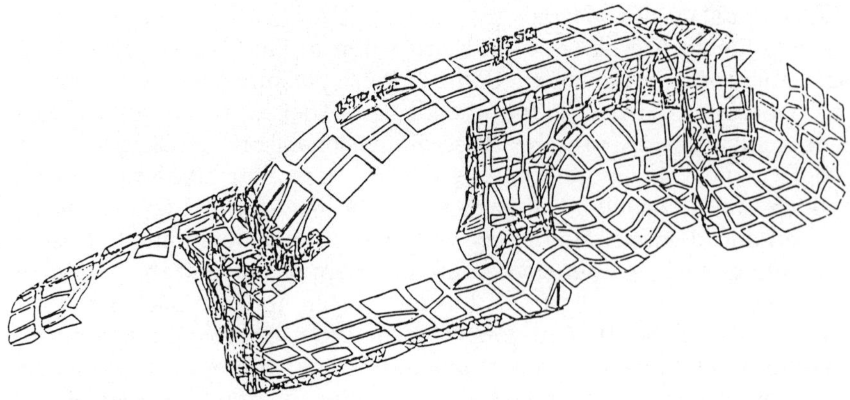

**Figure 9.10** Input data for structural analysis.

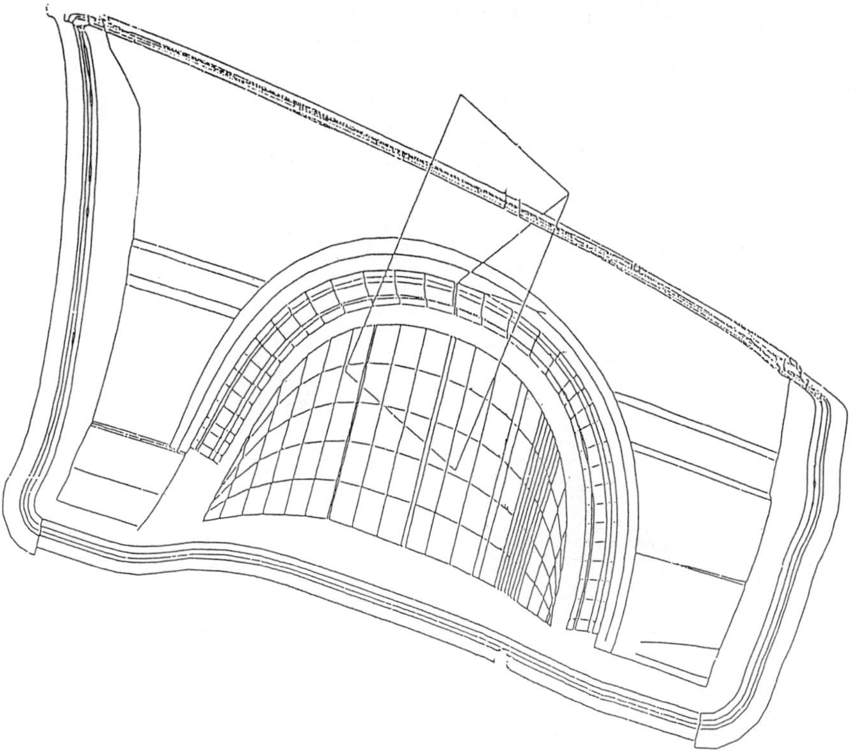

Figure 9.11 Surfaces of a stamping die.

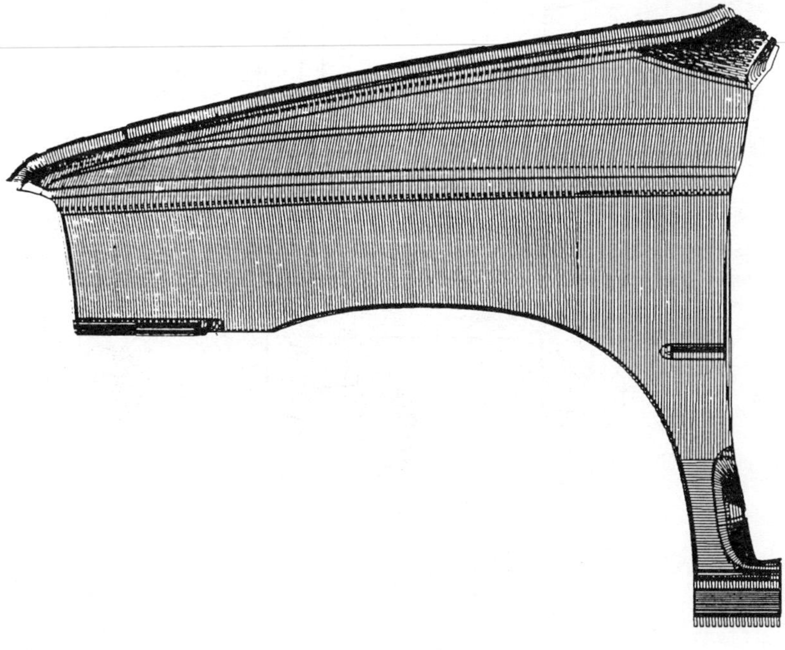

Figure 9.12 Cutter paths.

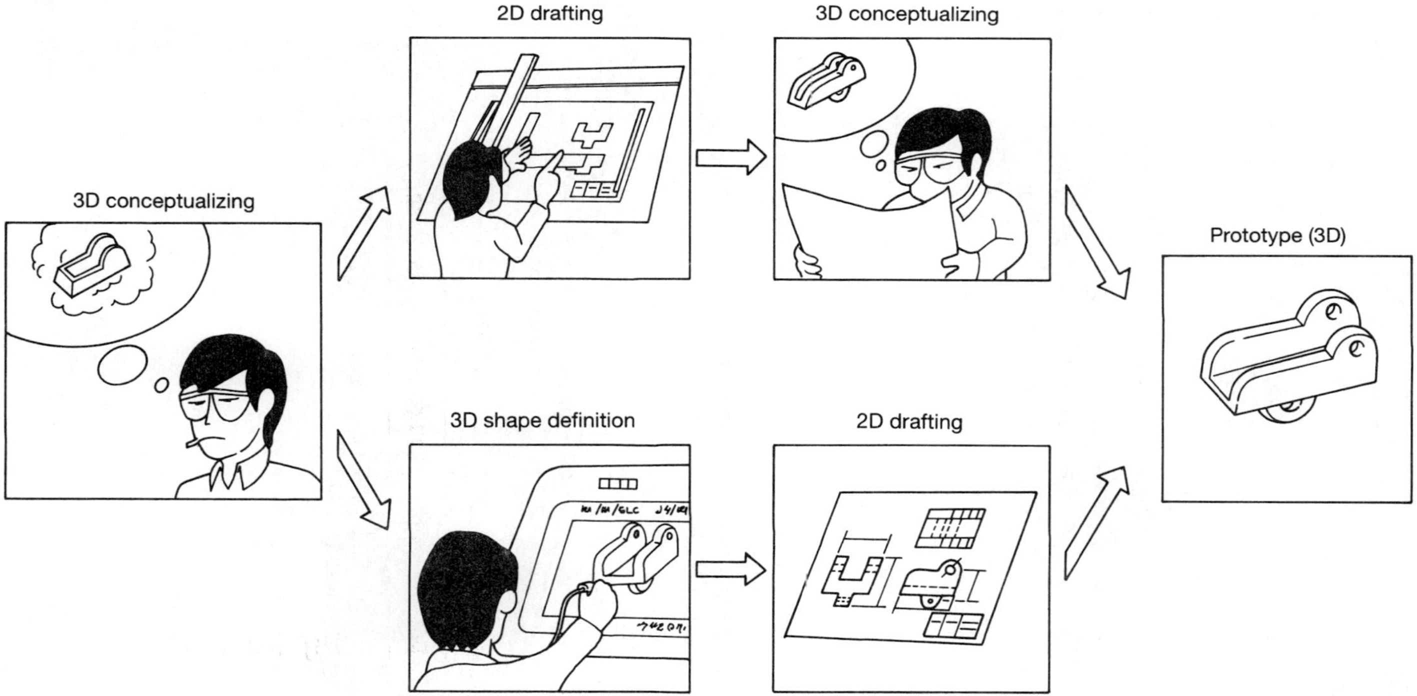


**Figure 9.13** Changes in design process.

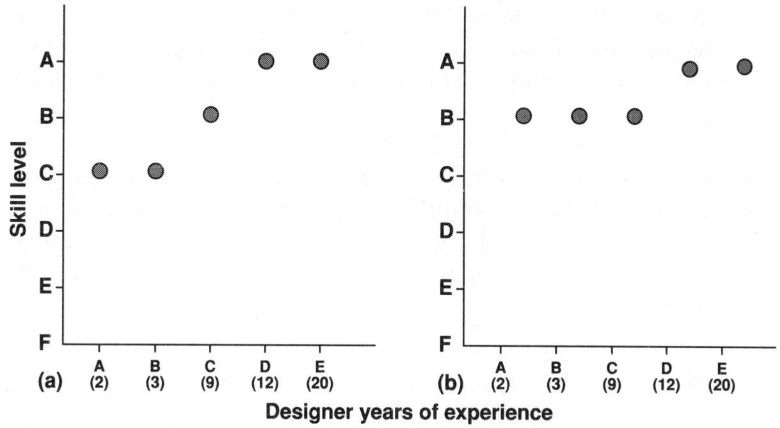


Figure 9.14 Reduction in skill-acquisition time: (a) by hand; (b) with GNC2.

One notable change which has stemmed from the full-scale application of the system is that younger people acquire design skills faster than with conventional procedures (Figure 9.14). Design skills such as sectioning and clearance checking traditionally required many years to acquire. These functions can easily be implemented using GNC2. The system facilitates the trial and error approach to rapid prototyping.

Changes have also taken place in the critical paths of the design and production preparation stages. The design of structure and shape can now be determined in the design stage and digitized concurrently as shown in Figure 9.15. Cutter interference, which is difficult to find in batch operation, can be processed by interactive operation in the

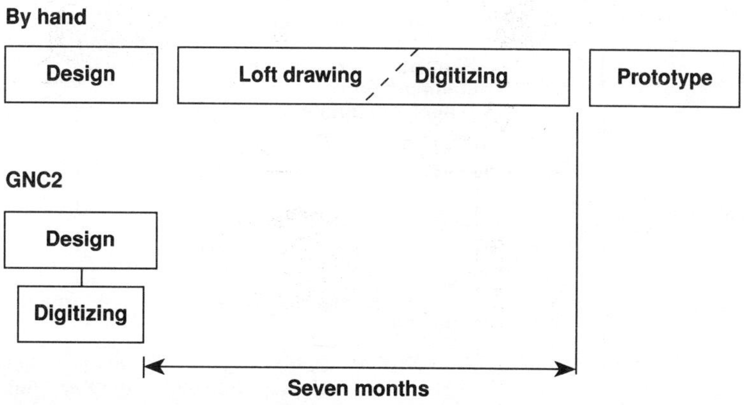


Figure 9.15 Change of critical path.

production preparation stage. This increases the NC cutting area and reduces the amount of hand finishing by transferring workloads to the later stages of development. This has the effect of reducing the total staff hours.

A five- to seven-fold increase in productivity over the manual design process can be achieved through shortened design times and reduce staff hours. This increase permits more frequent case studies which contribute to improved design quality. Improvements in development processes promote system operations and substantially reduce development times.

## 9.4 FUTURE SYSTEMS IMPLEMENTATION

The concept of a total system is shown in Figure 9.16. New areas where the system will be implemented are described in the following subsystems.

### 9.4.1 NON-BODY PARTS

GNC2 is intended for body surfaces. It is not, however, suitable for bulky components comprising primarily geometric surfaces, such as

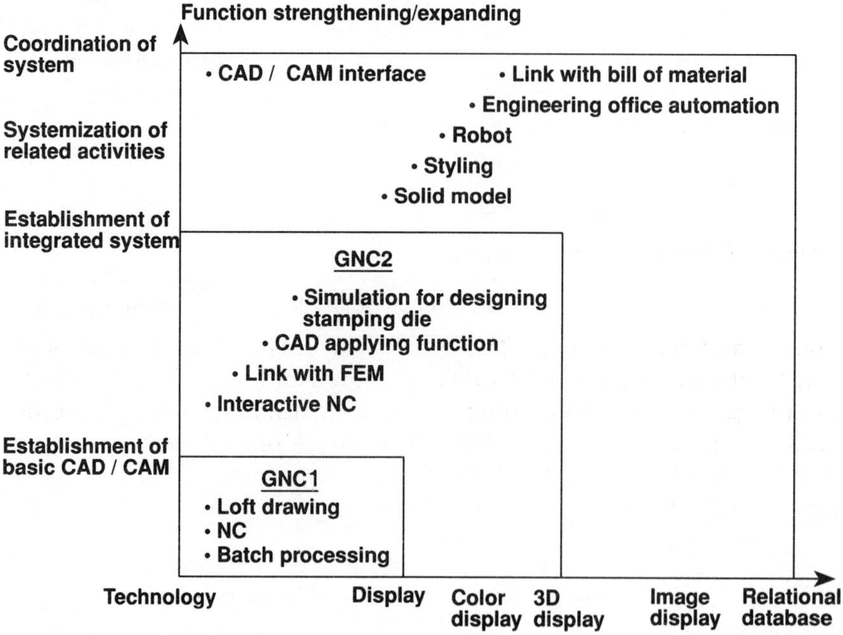


Figure 9.16 Development of system.

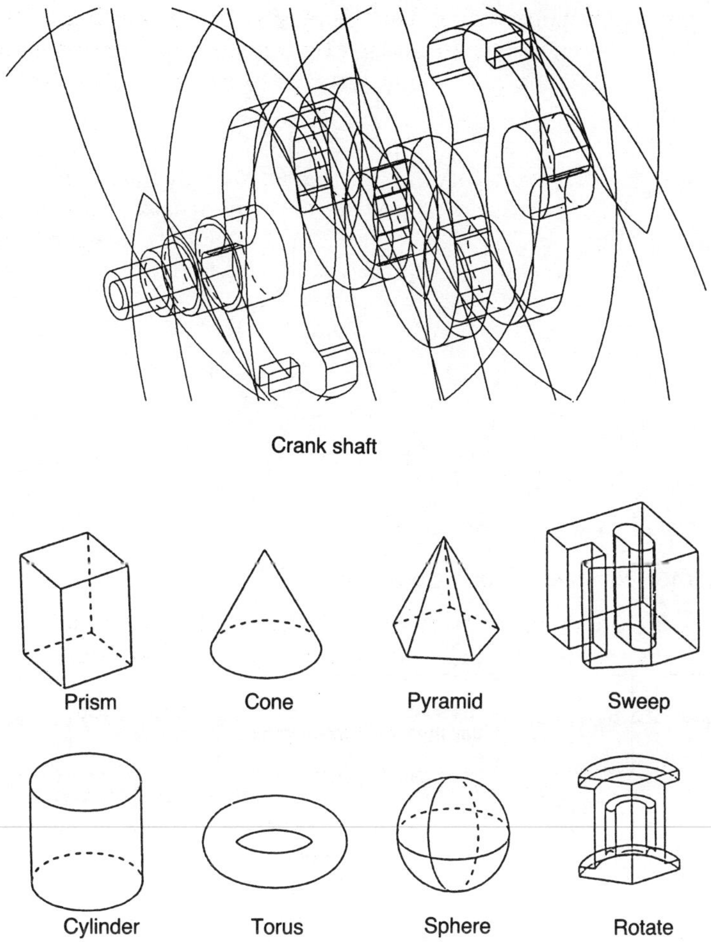


Figure 9.17 Primitive elements and connecting.

engines and transmissions. The reasons for this are considerations of shape definition time, shape processing time and CPU time. Development began on a prototype solid modeller, which is suited to geometric surfaces, in the autumn of 1982. The development of the primary functions has been completed. Solid modelling is now being applied to actual parts (Figures 9.17 and 9.18).

### 9.4.2 STYLE DESIGN

A parallel operation, shown in Figure 9.19, applies GNC2 to style design. An image processing system is also under development which

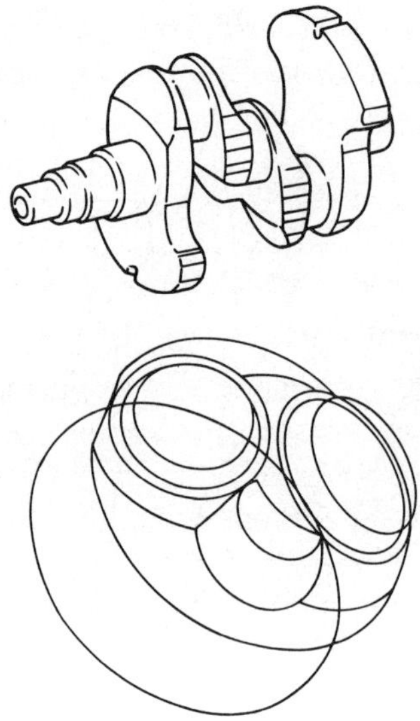

Figure 9.18 Examples of solid modelling.

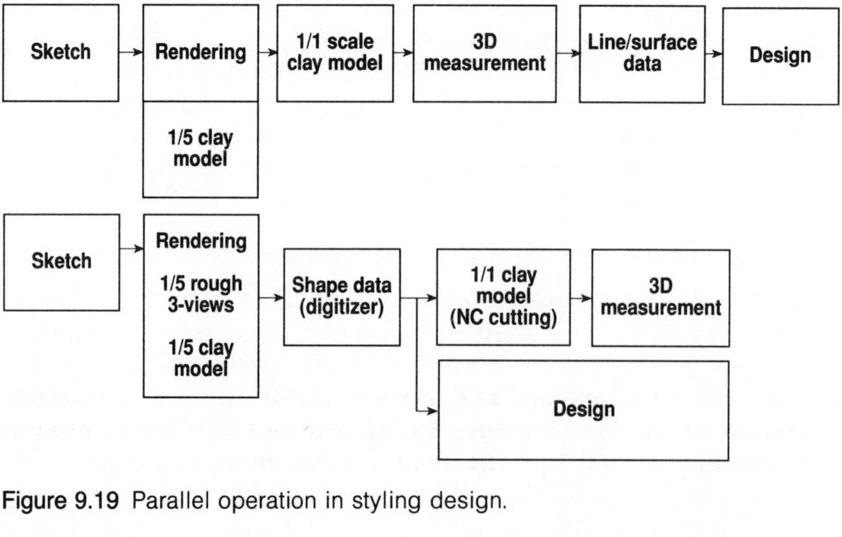


Figure 9.19 Parallel operation in styling design.

allows the description of three-dimensional shape data which can be displayed as a lifelike colour-shaded image for evaluation and review. The development of this system is complete and it is now being tested on actual development work. This system will shorten the time taken to perfect shape data in the design process.

### 9.4.3 BASIC DESIGN

Although GNC2 has so far been applied primarily to shape processing, there is an increasing need to integrate it with CAE. The structural analysis input/output function is being strengthened to meet this need.

### 9.4.4 DETAIL DESIGN

CAD has become widely accepted due to the increased use of GNC2 in design work. Design procedures are now evolving into automatic design procedures. An all-encompassing engineering system is also needed which is linked to the basic product information system and the technical information system.

### 9.4.5 PRODUCTION PREPARATION

A system is needed to reduce the time required for testing stamping dies, plastic moulds, and casting patterns and dies. To achieve this, a simulation program for contouring stamping dies is almost completed. There is also a strong demand for robot off-line programming via an interactive operation system.

### 9.4.6 CAD/CAM INTERFACE

The expanding use of CAD/CAM systems throughout the world is increasing pressure to replace hard-copy drawings with digital design information. This digital information takes the form of shape data and NC data. It can be delivered by means of magnetic media or by computer networks. Thus the information format and its transmission will need to be selected on the basis of each individual supplier's needs.

## 9.5 CONCLUSIONS

An overview of Mazda's CAD/CAM system and its future developments has been presented. Hardware functionality and performance will be improved, reducing operating costs. Integrated software systems will become a reality. CAD/CAM systems will continue to increase in both applications and number. Future efficient use of CAD/CAM requires new design procedures. CAD/CAM systems are no longer just a convenient tool, but are indispensable instruments for competitive and economic production.

# CIM in computer assembly

# 10

*Hioshi Watabe*

This chapter gives an overview of the approach to CIM at the IBM Fujisawa plant.

## 10.1 IBM AND ITS GLOBAL SYSTEMS

IBM is a multinational company which provides computer systems and systems solutions in more than 130 countries. It develops products at more than 30 development laboratories in the USA, Europe, Canada and Japan. Products are shipped from more than 30 plants located all over the world. Product development is carried out at designated laboratories, but production is often shared between plants. For example, the high-end processor systems are developed in the USA and manufactured in the USA, France and Japan.

Products are also ordered and shipped between these plants. For example, semiconductor plants ship their products to circuit packaging plants. These in turn receive orders from box (final customer goods) assembly plants.

IBM's operation is supported by global computer systems (Figure 10.1), in which the Fujisawa plant's CIM is designed to fit. Two kinds of common systems are included in the global system. Centre systems are included for optimization at the total enterprise level. Distributed

*Manufacturing Automation Systems and CIM Factories*. Edited by K. Asai, S. Takashima and P.R. Edwards. Published in 1994 by Chapman & Hall, London. ISBN 0 412 48230 4

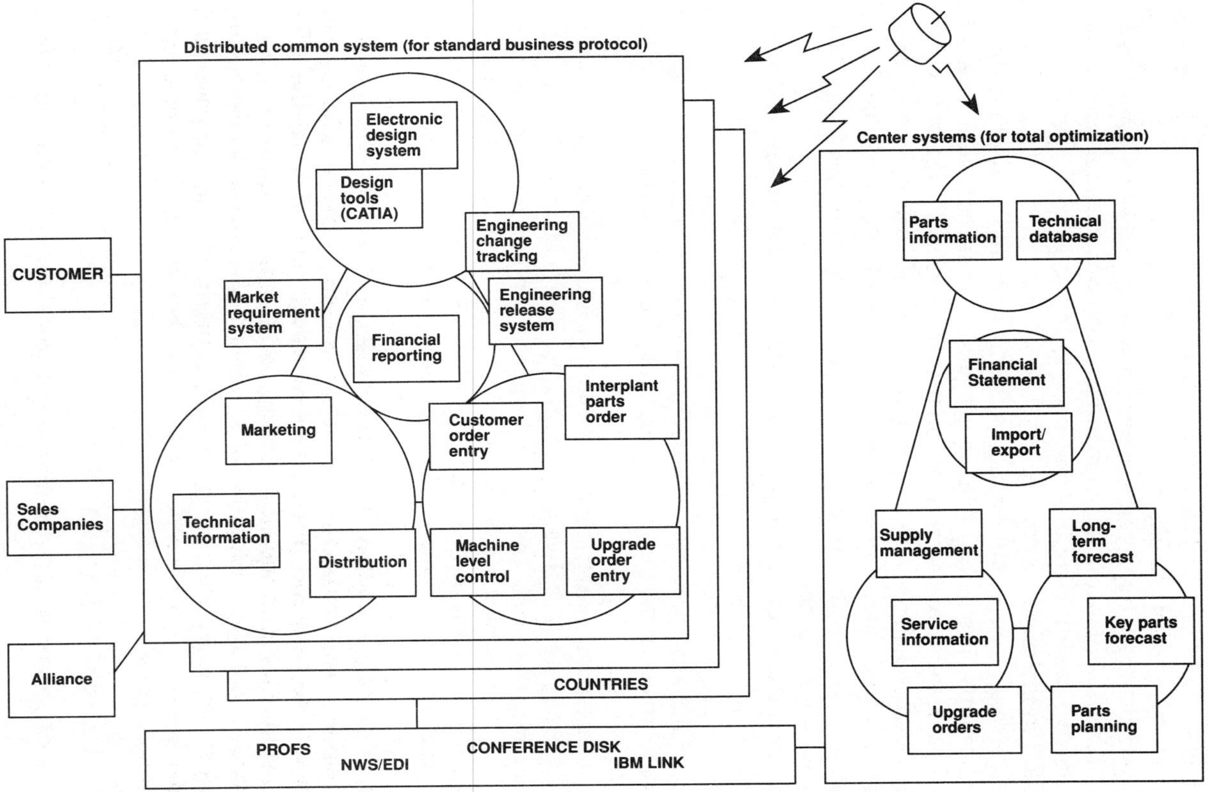


**Figure 10.1** Systems for IBM multinational operation. Fujisawa CIM is designed to work with these global systems.

common systems are included for standardization of business protocols between countries and/or operating units.

Through the marketing and supply management systems, orders from customers in many countries are received and checked for consistency of system configuration and features. Orders are sent to the central supply management system which establishes an optimum manufacturing plan and sends manufacturing orders to relevant plants. Shipment schedules are also fed back to the sales representatives according to the plant responses.

In the engineering information system, drawings and other design specifications, process specifications, and test specifications are sent from laboratories to the relevant plants through a standardized interface. Design changes (engineering changes) are also delivered in the same way and their implementation is traced by the system. Central databases for technical information and IBM parts provide on-line information for engineers.

In the production planning and parts planning systems, demand forecasts are broken down for each plant on a country basis. MRP (material requirement planning), based on this forecast, runs at a central MRP facility. It allows coherent operation between final assembly plants, sub-assembly (such as electronic circuit) plants, and semiconductor plants.

The interplant parts logistics system provides a standard protocol for ordering parts between IBM plants. The network also includes procurement centres. The physical distribution system manages international and domestic movement of IBM products and parts for customer delivery, production and service.

Person-to-person international electronic communication is established via office systems between all IBM offices. An individual can access his or her mailbox even from abroad. The network transmits data, documents, business graphics, and drawings. It also has an electronic bulletin board for company news, and databases such as a worldwide IBM employee telephone directory. Professionals and engineers exchange ideas using numerous conference boards in the network. This system is an important part of the IBM business infrastructure.

## 10.2 THE IBM FUJISAWA PLANT

IBM has two plants in Japan, one at Fujisawa, about 50 kilometres from Tokyo, and another at Yasu, about 30 kilometres from Kyoto. They ship IBM products to Asia Pacific countries. They also ship parts to other IBM plants in many countries.

Yasu produces semiconductors, electronic components, and high-end processors. Fujisawa obtains its electronic components from Yasu. The products of the Fujisawa plant can be divided into three groups, each of which requires different manufacturing technologies.

The first group, build-to-order products, are middle-range processors, high-end DASD (direct access storage devices) and their controllers, telecommunications controllers, and system printers. These products offer many customer options. Machines are assembled and tested according to the specification of each order. The important aspect in manufacturing is the logistics of meeting or changing orders quickly with the minimum material inventory.

The second group, build-to-plan products, include personal computers and table-top printers. These are assembled and shipped from bulk orders from the marketing team. Manufacturing criteria are low-cost manufacture, quality control, and flexibility for dynamic demand changes. These products also tend to have many types and models, even those based upon a single model. The capability to manage the production of mixed models and types is also required.

The third group is technology components. The Fujisawa plant is one of the biggest suppliers of 3.5 inch and 2.5 inch hard discs in the world. These discs are used in personal computers and workstations. Although they constitute a build-to-plan product, the high production volume requires further manufacturing technology such as automation and real-time monitoring and control of manufacturing lines.

## 10.3 CIM AND IBM

IBM is pursuing CIM as a company-wide strategy, to improve its own productivity and to provide CIM solutions to its manufacturing customers. The company motto is 'use what we sell, sell what we use'.

The computer is an important source of competitiveness for a company, as illustrated by the famous success stories of strategic information systems, such as the one at American Airlines. It is not yet easy to integrate computers in the complex environment of a manufacturing company, where many islands of automation and computer systems exist. There are large amounts of data associated with CAD and big business databases, and a proliferation of automation controllers.

IBM is committed to providing solutions in this environment. In 1989 it announced CIM architecture as the basic framework for integration, and is committed to providing CIM products. IBM has also implemented company-wide management systems for the promotion of CIM in its manufacturing facilities and the development of CIM hardware and software. This is useful for both internal plants and customers.

## 10.4 CIM STRATEGY AT THE IBM FUJISAWA PLANT

As for any other company, the Fujisawa plant is always faced with keen competition. It needs to be better than its sister plants within IBM, otherwise the sales force may sell goods imported from other plants. It must be competitive in terms of cost and quality for low-end hard disc drives in order to maintain its position of supplier to the personal computer plant.

The plant recognizes CIM as an important means of maintaining competitiveness and versatility. It aims to become an advanced CIM plant by taking advantage of being a computer manufacturer. Its approach to CIM can be summarized as follows:

(a) It should be based upon IBM's CIM concept.
(b) CIM must be promoted together with operational improvements and automation, and the integration of these activities.
(c) It should combine the advantages of the IBM culture (architectural approach) and Japanese industrial culture (cumulative improvements).
(d) The plant should practise 'use what we sell, sell what we use'.
(e) It should pursue integration.
(f) CIM strategy should be reviewed periodically and its consistency with plant strategies confirmed.
(g) CIM must be included in new product plans.
(h) CIM should be built to fit in well with IBM's global systems.
(i) The auditability and security of systems must be ensured.

The plant develops five-year plans on an annual basis. These plans mainly include objectives and the necessary conditions to beat the competition, rather than the 'nice to have' challenging management objectives. CIM plans are developed in line with plant strategy.

## 10.5 FUJISAWA PLANT CIM

The application architecture of the Fujisawa CIM is shown in Figure 10.2. The applications are in five layers:

(a) **Management control layer**. This layer supports the plan and actual data for key measurement items and decision support systems. The actual data comes from layer two.
(b) **Plant administration layer**. This layer has two items: the interface to the IBM global systems described in section 10.2; and plant administration systems such as finance and personnel systems. This layer has an integrated database containing a summary of plant operations.
(c) **Engineering information systems and material logistics systems**.

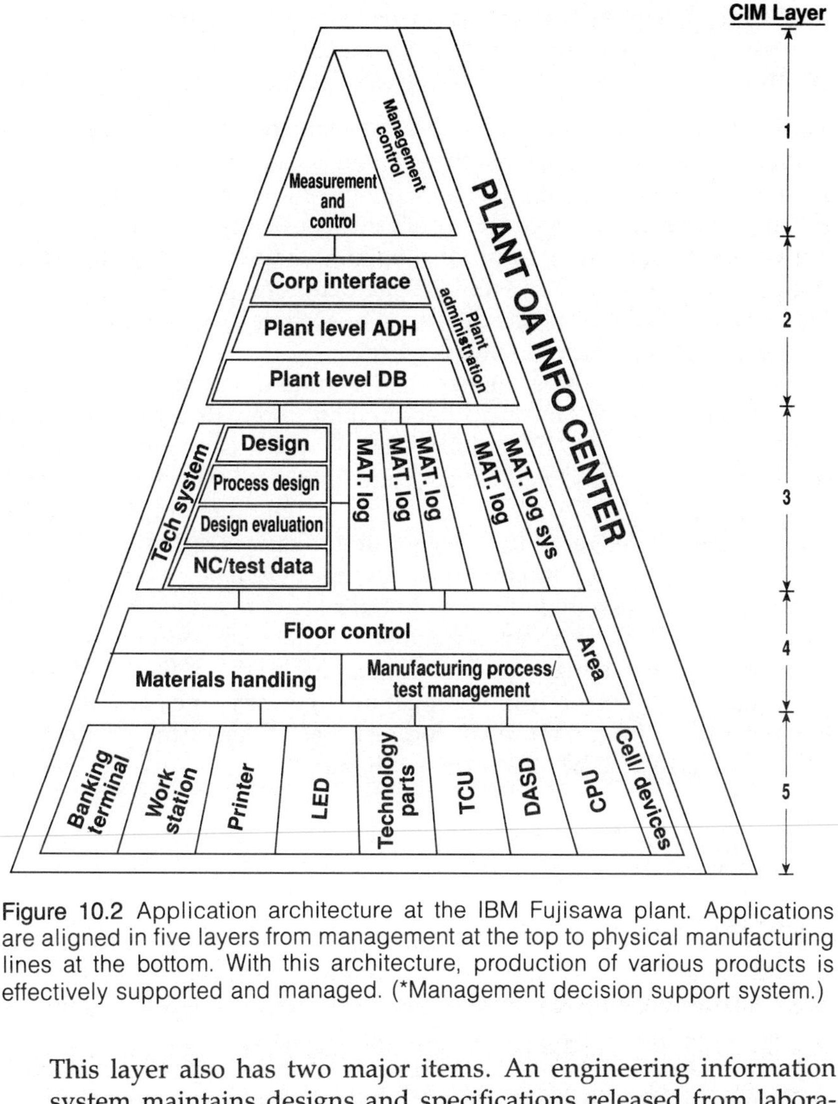


Figure 10.2 Application architecture at the IBM Fujisawa plant. Applications are aligned in five layers from management at the top to physical manufacturing lines at the bottom. With this architecture, production of various products is effectively supported and managed. (*Management decision support system.)

This layer also has two major items. An engineering information system maintains designs and specifications released from laboratories. This includes CAD and other facilities for plant engineers. Material logistic systems are for production planning, material planning and ordering, shipping, and inventory control. The vertical lines in the figure show several material logistics operation units according to production types. A unit can operate independently from other groups using the same systems.

(d) **Area (manufacturing floor) control systems**. These are the systems for monitoring and controlling a production line and for materials handling.

(e) **Physical manufacturing lines**. This layer includes robot controllers, testers, NC machines, and other automated equipment. Most of these are connected by LAN between each other and to layer four.

These layers are supplemented by the plant office automation information centre. This supports office systems, databases for office automation, and consulting services.

The layer concept keeps the application structure simple. The alternative is a need for time of product specific definitions for applications with different scope. This would jeopardize total performance. At the Fujisawa plant, the products described in section 10.3 are supported by this single CIM application architecture.

For example, take the example of 3.5 inch hard disc manufacture, the main product line at Fujisawa. Requirement forecasts and bulk orders are received from the IBM global system. MRP is performed, based upon the master production plan, and parts orders are placed with suppliers through EDI. Parts are also ordered from the Yasu plant. Major parts are delivered to the plant under kanban control and stored in the automated warehouse for a short time before being automatically delivered to the assembly line by another kanban signal. Products are assembled in a clean room on a line with a high automation ratio. Fully automated testing, tracking and logging of the individual units is performed. Any person from foreman or engineer to the plant manager can determine, from a terminal, the real-time status of the production.

The Fujisawa CIM is further characterized by five levels of integration:

(a) **Vertical integration**. The control data is passed from upper layers to production lines. Production status is passed from layer five to layer one.
(b) **Horizontal integration**. The same systems are used, with minimal differences, for different types of products (see section 10.3).
(c) **Integration with global systems**. The Fujisawa CIM is integrated with the IBM global system through layer two.
(d) **Intergrated database**. A database is provided for end-user computing. Users are able to use data through their terminals for analysis and planning.
(e) **Network integration**. The terminals and systems are connected by the SNA network. A user can access any application, any data, and any IBM location in the world, after proper authorization.

## 10.6 BENEFITS OF CIM AND THE FUTURE

Over a three-year period, the integrated efforts in CIM, automation and operation improvement resulted in a productivity improvement of

50%. Manufacturing lead times were reduced by almost 60% and inventory turnover by nearly 50%. The plant survived rapid currency fluctuations and managed to increase its competitiveness, the primary goal of CIM.

CIM will never be complete as long as competition exists. Competition necessitates improvements for survival. CIM emphasis will move to areas such as faster production, better quality, better operational integration with suppliers, better process control, and dynamic scheduling to maximize output.

IBM's CIM products, and the products of IBM's partners using the IBM CIM architecture, will gradually allow the systems to be built and changed more easily, more quickly and more cheaply. Systems should be built with a long-range architecture. The new CIM products and architecture will make it possible to focus more on operational improvements, by making system implementations easier.

# Robots in mass-production automatic assembly

# 11

*Toshijiro Ohashi, Seii Miyakawa and Makoto Matsunaga*

Trends toward greater consumer choice and shorter product life-cycles have made it essential for automatic mass-production assembly lines to be very flexible. They need to be easily adaptable to product changes, cheaper to build, and quickly set up. Recent improvements in robot technology have led many to believe that robots are the key to such flexible automatic assembly. Although it is likely that robots will fill this role in future systems, few robots have yet been used in mass-production facilities as they are still relatively slow and expensive.

A systematic approach, from product design improvement to facility development, is necessary for developing flexible automatic assembly lines which effectively use robots. This chapter describes product design improvements, and computer simulation for forecasting line performance. It also outlines an assembly line using robots for flexibility, and shows the specifics of robot utilization.

## 11.1 PRODUCT DESIGN IMPROVEMENT

Product design plays an important role in the automation of assembly, product life, and product quality. Products which are difficult to assemble require more sophisticated, dextrous, and hence expensive assembly facilities. More complex assembly stations have a higher failure rate and reduce the performance of the assembly line. The authors of this chapter have thoroughly reviewed product design and have developed a new VCR mechanism which is suitable for automatic

*Manufacturing Automation Systems and CIM Factories*. Edited by K. Asai, S. Takashima and P.R. Edwards. Published in 1994 by Chapman & Hall, London. ISBN 0 412 48230 4

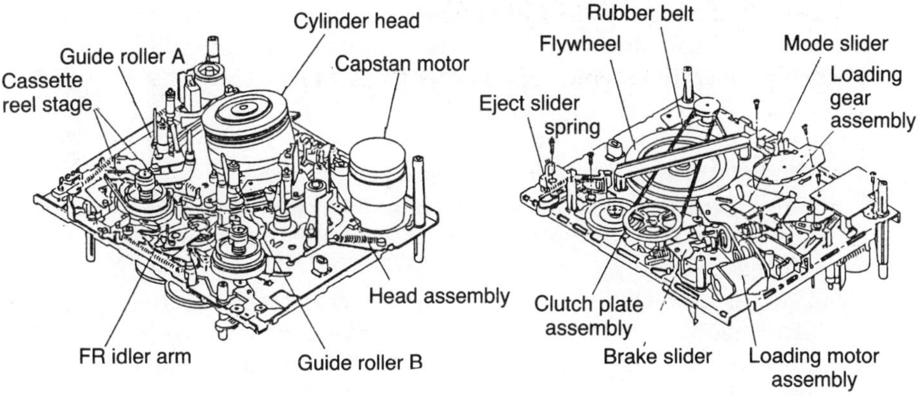


Figure 11.1 VCR mechanism.

assembly. Figure 11.1 shows the mechanism which was developed. Most of the parts, including pressed parts, moulded parts, rubber belts, and coiled springs, are assembled above using a simple attachment motion. The result of the design improvement is that these parts can be attached by pick-and-place units or relatively inexpensive assembly robots.

The product design review used an analytic design improvement procedure called the assemblability evaluation method (AEM). This was developed at Hitachi (Awane *et al.*, 1981; Hashizume *et al.*, 1980). AEM analyses assembly structures using approximately 20 symbols which give designers and production engineers a quantitative measure of how easily products can be assembled. This analysis highlights weaknesses in product design (location and reason) in terms of assemblability (assembly producibility). Figure 11.2 shows the three basic features of AEM.

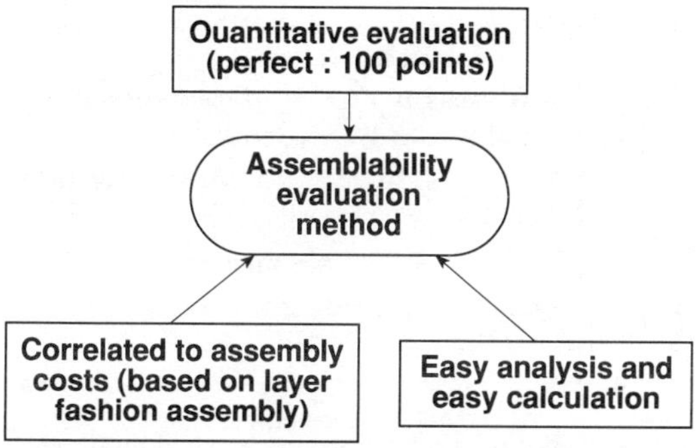


Figure 11.2 Features of Hitachi assemblability evaluation method (AEM).

Firstly, evaluation indices are quantified, based on a hundred point maximum. This allows an easy determination of the difficulty of assembly operations and easily identifies design features in need of improvement. The evaluation indices can be used as a common language for designers, production engineers, and managers. The indices also constitute an effective management tool.

Secondly, AEM is easy for designers to learn and use. This makes it possible to evaluate and improve the assemblability of a product in the early design stages.

Finally, assemblability evaluation indices are correlated to assembly cost. This allows the deduction of the standard time for assembly and assembly costs. Designers can then directly evaluate the effect of assemblability improvement in terms of cost.

Two indices are used in AEM: the assemblability evaluation score, which is an index of design quality; and the assemblability cost index. The assemblability evaluation score assigns a value of 100 points to the most easily assembled case. Points are deducted for elements which reduce assemblability.

Figure 11.3 shows that the estimated assembly operation costs, using normalized values, for various product types are very close to the actual production costs. This proves that this method is sufficiently accurate for practical purposes and that assembly operation costs can be easily and accurately estimated at an early design stage. Designers can effectively and efficiently improve product design during the early stages of development by means of the evaluation and improvement

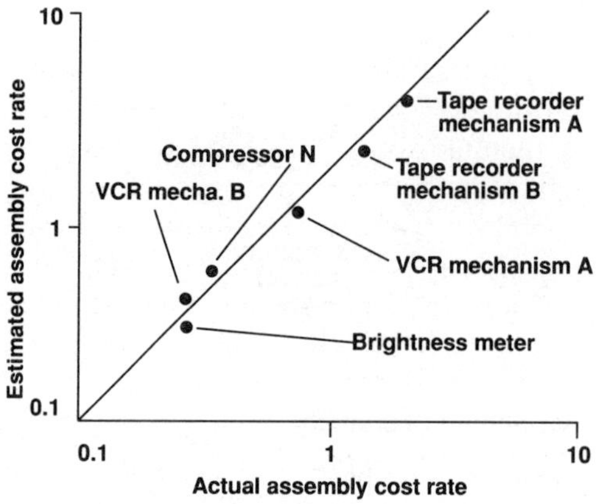


Figure 11.3 Accuracy of cost estimation by AEM.

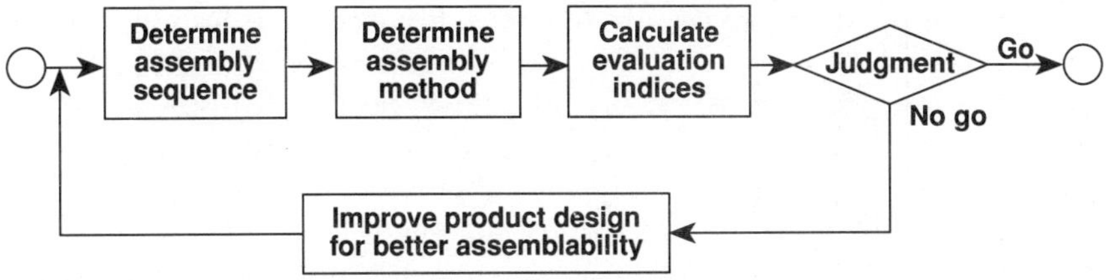


Figure 11.4 Design evaluation and improvement steps by AEM.

iterations shown in Figure 11.4. This method provides a number of advantages:

(a) facilitation of factory automation;
(b) reduction in assembly labour;
(c) shortened design periods;
(d) improved reliability of products and automated equipment.

Table 11.1 gives examples of design improvements in a VCR mechanism. The results of these improvements are summarized in Table 11.2.

## 11.2 LINE PERFORMANCE SIMULATION

In order to develop an automatic assembly line effectively, the authors examined the relationship between line specification and line performance during system design. An assembly-line simulation was developed using the general-purpose simulation system (GPSS), a simulation language based on queuing theory. Figure 11.5 shows the input and output data for the simulation.

After several trial simulations of the line, the buffer size and distribution were determined. The final simulation results are shown in Figure 11.6. These results showed that the line was capable of achieving the intended eight-second cycle time. These results were used to determine the assembly machine specifications. Table 11.3 compares the forecasted average line tact time, the actual tact time after six months' operation, and the predicted tact time using the actual station data. Although the average machine cycle time was longer than the value which was estimated prior to development, the line achieved target production because of a lower than average breakdown rate. The table also shows the accuracy of the simulator. The line tact time of 7.9 s, calculated from the actual station data, was very close to the actual tact time of 8.0 s.

Table 11.1 Examples of design improvements in the VCR mechanism

| | | | | | | |
|---|---|---|---|---|---|---|
| Purpose | Reduction of number of parts | Stabilize parts positioning | Easy handling | Easy insertion | Prevent entanglement | Prevent jamming |
| Means | Unitize parts | Prepare positioning guide portion | Prepare parallel portion for gripping | Prepare chamfering | Use connecter to eliminate lead wires | Increase parts thickness |
| Part name | Tape guide | Loading motor assembly | Capstan motor assembly | Screw | Loading motor assembly | Washer |
| Before design improvement | 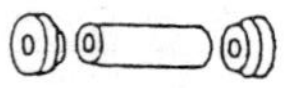 | 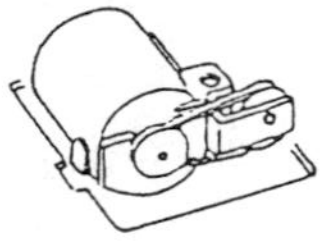 | 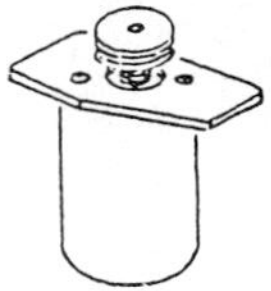 |  |  |  |
| After design improvement |  |  | 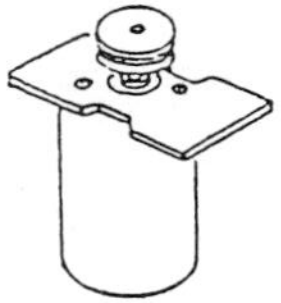 |  |  |  |

Table 11.2 Assemblability improvement result in VCR mechanism

| | New mechanism | Previous mechanism |
|---|---|---|
| Number of parts | 379 | 460 |
| Assemblability evaluation score | 73 | 63 |

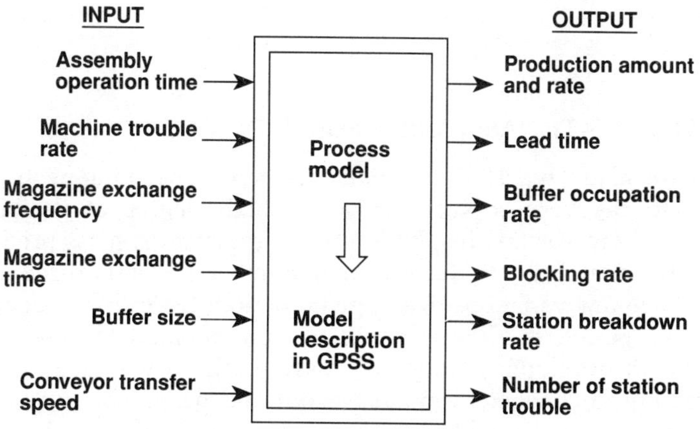


Figure 11.5 Input and output of the simulation program.

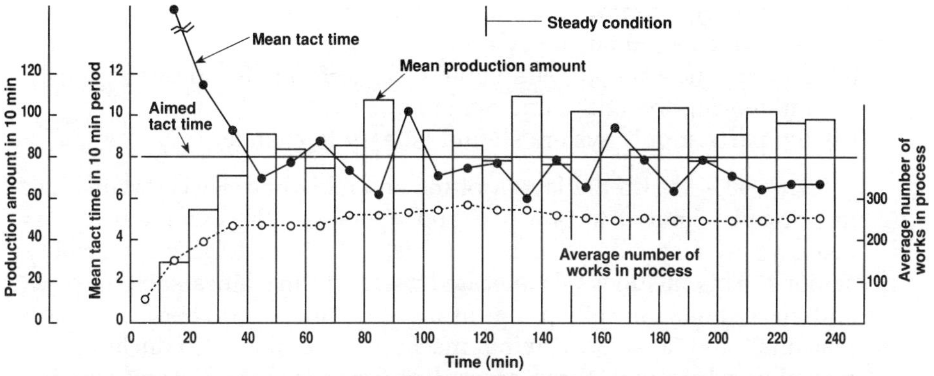


Figure 11.6 Line performance simulation result.

Table 11.3 Comparison between simulation result and actual line performance

| | Simulation (planning) | Actual production result* | Simulation (based on the actual data)* |
|---|---|---|---|
| Average machine cycle time (s) | 4.25 | 5.3 | (5.2) |
| Average machine breakdown rate (%) | 2.2 | 0.3 | (0.3) |
| Line tact time (s) | 7.5 | 8.0 | 7.9 |

* Data 6 months after the line operation started

## 11.3 AUTOMATIC ASSEMBLY FOR VCR MECHANISMS

There are many types of automatic assembly line, ranging from the simple to the very sophisticated. The high degree of flexibility is desirable, however this results in greater expense. A mass-production assembly line usually produces a small number of standardized models which have few variations over a given period, typically a year or two. After this period the line is modified to accommodate new models. The aim in developing automatic assembly lines was to incorporate adaptability to changes in model, layout and parts. This required the facility to make quick and easy changes in parts supply and assembly operations. The facilities also had to undergo minor modifications and be ready for production in a short time period, the reason for this being that market demands usually require new models to be available in large quantities and at short notice. In view of these demands, the assembly line was designed to satisfy the following conditions:

(a) individual assembly machines should be simple;
(b) the reprogramming of each station should be easy, and should be performed simultaneously on all stations;
(c) maintenance and repairs should be performed without seriously influencing the entire line operation;
(d) the parts supply system should be easy to change.

Figure 11.7 shows the layout of the assembly line. Each station was, in principle, limited to one part and operation, the exception to this being screw fastening. The line consists of 86 modularized assembly stations. This includes 56 dedicated assembly machines such as pick-and-place units, a coil-spring fitting machine, a rubber-belt fitting machine, and a screw-driving machine. The line also includes 11 assembly robots with three degrees of freedom. Inexpensive, variable stroke pneumatic pick-and-place units are used for chassis loading

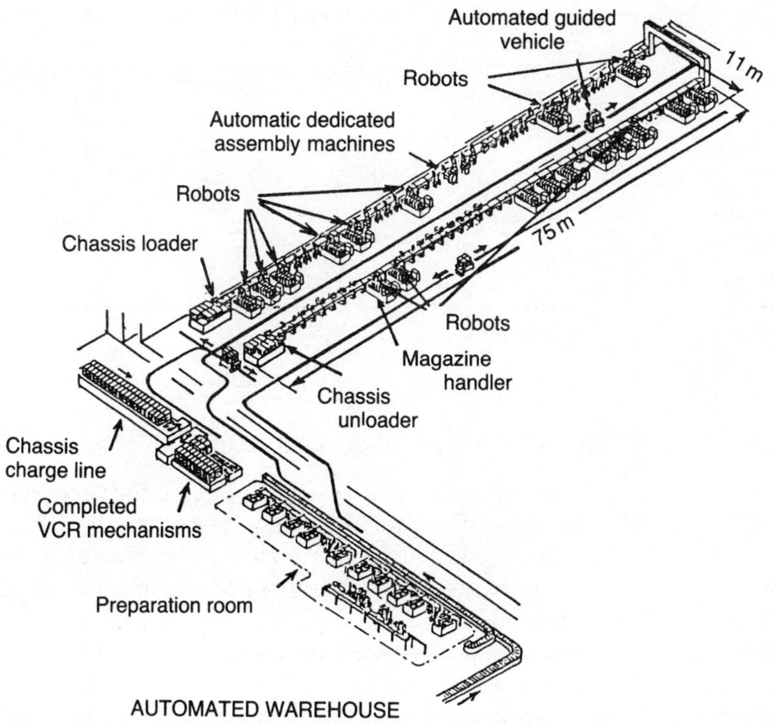


Figure 11.7 Layout of automatic assembly line for VCR mechanisms. Stations: 56 automatic dedicated machines, 11 robots, 9 manual assembly, 10 buffers – total 56. Automation rate (= number of automated stations ÷ number of assembly stations × 100) = 88%. Tact time 8 s.

and unloading operations. Sub-assemblies which require complicated attachment operations are attached manually at nine stations. The chassis, assembly parts, and sub-assemblies are contained in flat magazines or buckets. They are supplied by automatic guided vehicles from automated warehouses and preparation rooms to each station. Production is controlled by a Hitachi L-320 minicomputer and a H-08 process control computer. The mean tact time of the line achieved the target of eight seconds. A view of the line is shown in Figure 11.8.

## 11.4 CONSTRUCTION OF AUTOMATIC ASSEMBLY STATIONS

Figure 11.9 is a block diagram showing the construction of the assembly stations. The facilities are standardized using block-built units. The standardized units can be used even when major model changes are required. Robots and basic machines are especially adaptable to

Figure 11.8 Automatic assembly line for VCR mechanisms.

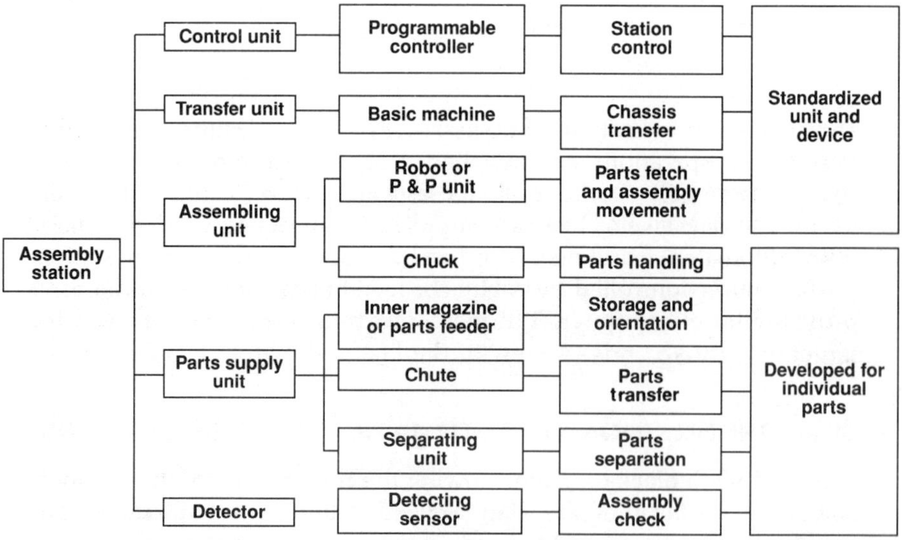


Figure 11.9 Block diagram of assembly station.

model changes which only involve program changes and machinery adjustments.

Other units which were developed to handle specific parts may or may not be adaptable for new models. If not, they can be replaced with new units. This is a viable option due to their relative low cost.

Figure 11.10 shows a robot station. It consists of the basic machine, an A3020 assembly robot developed by Hitachi, and a magazine handler. The basic machine is of a modular, direct-feed, free-cycle non-synchronous type. Although the direct-feed feature requires the designers to prepare supporting legs and positioning features on a chassis, it does not require a large number of pallets which are expensive to make. In addition, the line can accommodate more chassis between stations at a time, thus increasing productivity. The modular structure allows layout changes, readily allowing the number of stations to be increased with rising production output. Individual stations can also be simultaneously developed and adjusted at various sites by different manufacturers, permitting rapid line setup. The robot station does not need an x-y table for magazine positioning, allowing the magazine handler to be smaller and cheaper.

The robots are used for delicate and precise sub-assemblies or complex shaped parts supplied in flat magazines. Owing to design improvements, the robots on this line can be relatively simple and inexpensive. Sub-assemblies which are to be attached by robot were selected based on AEM analysis of the attachment operations. Some of the sub-assemblies or parts require intricate attachment operations that make full use of robot functional flexibility. Although the producibility design of the VCR was improved, some sub-assemblies or parts require intricate attachment operations and must be assembled manually. Most of these operations could be performed by high-performance robots. However, it was decided not to adopt such sophisticated robots and to perform these operations manually instead. This helped to maintain the cost performance of the line as well as to allow faster development of the line. Figure 11.11 shows the A3020 assembly robot developed at Hitachi.

The completion of the automated assembly line for VCR mechanisms provided the following benefits for eight-hour operation:

(a) assembly labour was reduced by a factor of six from 180 to 30 workers;
(b) the assembly floor area required was reduced by a factor of three;
(c) the quality improvement of the readjustment rate of final assemblies was reduced by a factor of five.

Line performance has been improved since completion of the line. Today, the line is operating 16 hours per day at an average line tact time of four seconds, quadruple the value of the benefits described.

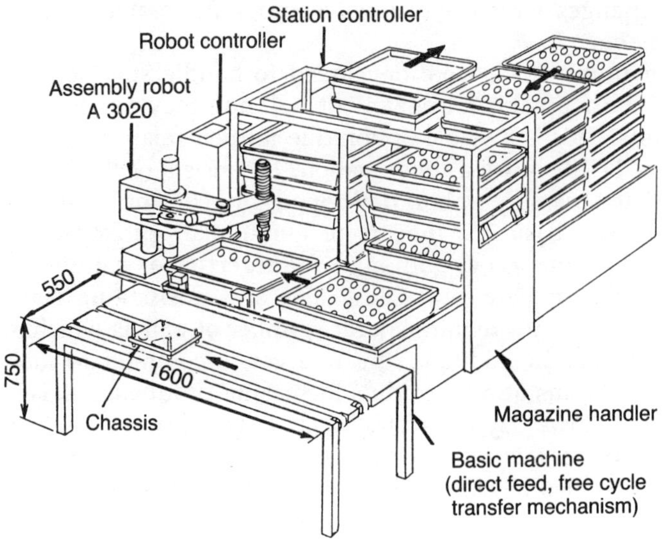


Figure 11.10 Robot assembly stations.

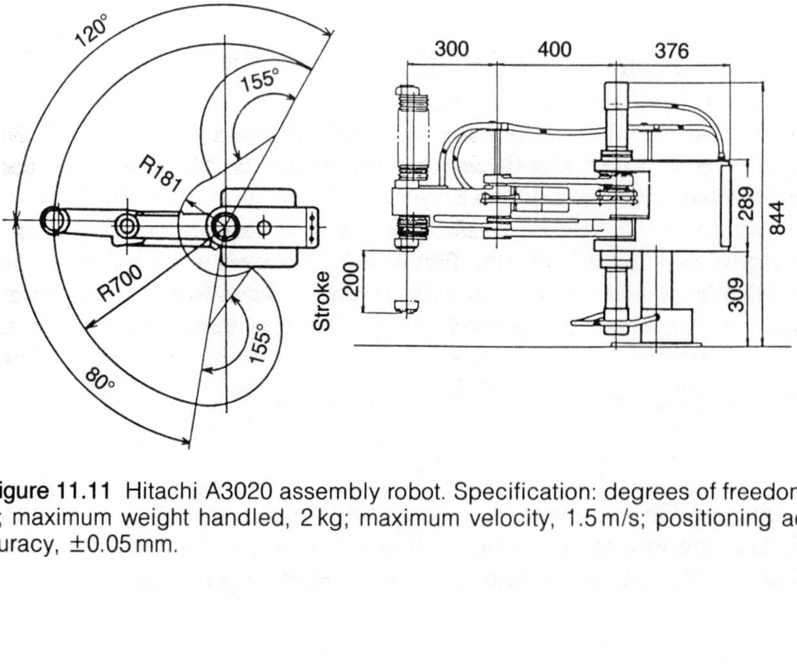


Figure 11.11 Hitachi A3020 assembly robot. Specification: degrees of freedom, 3; maximum weight handled, 2 kg; maximum velocity, 1.5 m/s; positioning accuracy, ±0.05 mm.

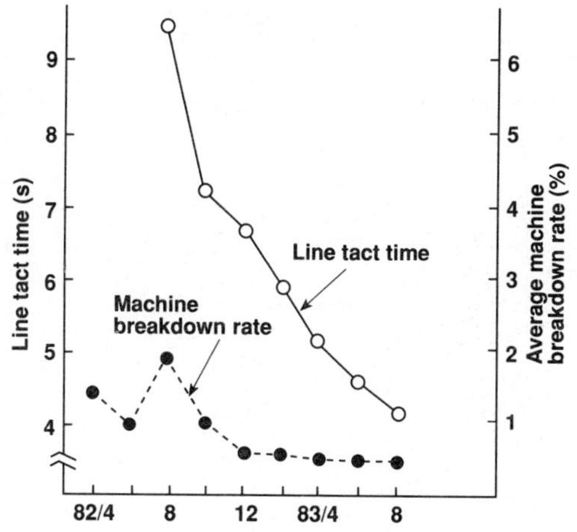


Figure 11.12 Improvement in machine breakdown rate and line tact time.

## 11.5 IMPROVEMENT OF LINE PERFORMANCE

After the development of this automatic assembly line, it became necessary to greatly increase the line's output to meet successive increases in VCR production targets. A number of steps were taken to achieve this, including:

(a) reducing the machine breakdown rate at the automatic assembly stations through failure analysis and preventative maintenance;
(b) improving the parts feeding systems;
(c) using temporary backup personnel in the maintenance of bottleneck stations.

These measures cut the line tact time to four seconds, half of the initial time. Figure 11.12 shows the improvements in the average station breakdown rate and line tact time.

Using a VCR, station breakdown analysis examined the causes of machine operation failures. Figure 11.13 shows the results of line simulation results based on current data. Simulation results give a tact time of about 5.6 seconds. This suggests that the effect of the temporary backup personnel in the maintenance of bottleneck stations is to reduce the average tact time by 1.6 seconds.

A 100% automated assembly line, based on this assembly line, was developed in 1986 to meet the expanding market requirement. A new mechanism was developed in 1987, reducing the number of parts from 379 to 250. The assemblability score was improved from 73 to 78. The first assembly line was restructured for this new model, and a 100% automated and faster assembly line with a mean tact time of 3.5

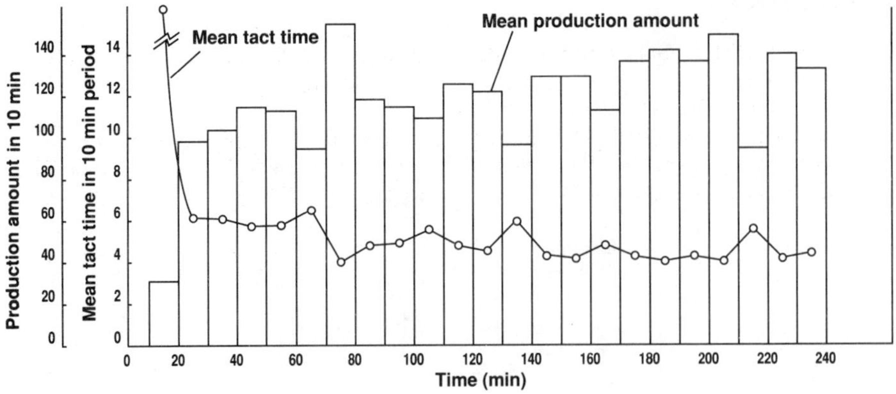


Figure 11.13 Line performance simulation result (after improvement).

seconds was completed. As a result of these developments and improvements, these two assembly lines had produced 16 million VCR mechanisms by the end of 1988.

## 11.6 CONCLUSIONS

The development of an automated assembly line incorporating robots for the assembly of VCR mechanisms has been shown, and the line and its stations have been described. The use of assembly robots shows great promise for the further development of such assembly lines. Further progress in this direction rests upon solutions in the following areas:

(a) development of assembly robots with better cost performance;
(b) preparation of a full range of menus for function, accuracy, and other factors;
(c) incorporation of artificial intelligence functions such as machine vision;
(d) greater adaptability of grippers and chucks, and the development of practical automatic hand changers;
(e) improvement in failure rate monitoring and operation monitoring, and the development of an auto-recovery function.

As progress continues in factory automation, the goal is a greater general-purpose utilization of assembly robots, and the achievement of both higher precision and greater functional sophistication.

## REFERENCES

Awane, H. *et al.* (1981) Computer aided planning of a fully automated assembly system for tape recorder mechanisms. *Proceedings of AUTOFACT*, 9–12 November, Detroit, USA, pp. 487–497.

Hashizume, S. *et al.* (1980) Development of automatic assembly system for tape recorder mechanisms. *Research and Development in Japan Award*, the Okochi Memorial Prize, Okochi Memorial Foundation, pp. 52–56.

# Robot spot-welding 12

*Junichi Kawamura*

Robots used in welding may be broadly classified as either spot-welding robots or arc-welding robots. Spot-welding robots, as shown in Figure 12.1, were the first to be developed. Arc-welding robots were introduced several years later. Despite this fact, arc-welding robots greatly outnumber spot-welding robots. Spot-welding robots first appeared in Japan around 1968 in the form of the American-made Unimate. The history of arc-welding robots began with the introduction of the Motoman by Yaskawa Electric, ten years later in 1977.

One major difference between spot-welding and arc-welding robots is their payloads. An arc-welding robot has a relatively small payload, three to five kilograms, in the form of its torch. A spot-welding robot has a payload of 60 to 100 kilograms, due to its heavy spot-welding gun. Similarly the positional repeatability of a spot-welding robot is low (0.5–1 mm) compared to the more precise (0.2–0.3 mm) arc-welding robots. In the early days, when the Unimate was developed, the power source was an electro-hydraulic servo, which matched the spot-welding functional requirements. This power source was, however, unable to meet the positional accuracy requirements of arc-welding robots. Developments in electric servo-motors and control technology have made it possible to achieve the accuracy requirements of arc-welding robots. This made the Motoman realizable.

Spot-welding robots are also gradually changing over to electric servo-motor drives due to the shortcomings of hydraulic systems, for example maintenance problems, warm-up time requirements, and low efficiency. Figure 12.2 clearly shows that the largest demand for spot-

*Manufacturing Automation Systems and CIM Factories*. Edited by K. Asai, S. Takashima and P.R. Edwards. Published in 1994 by Chapman & Hall, London. ISBN 0 412 48230 4

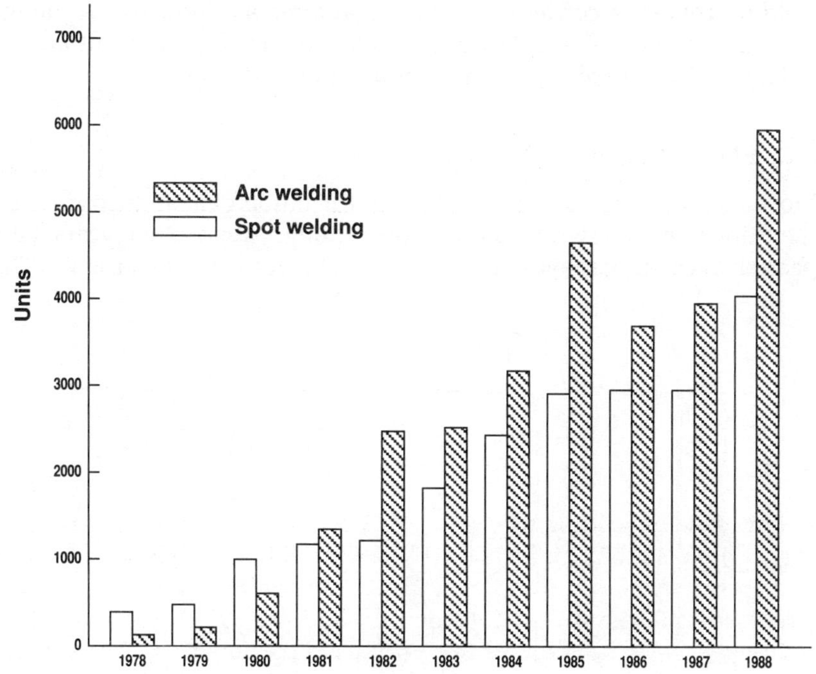


**Figure 12.1** Growth of welding robots in Japan (playback, NC and intelligent robots).

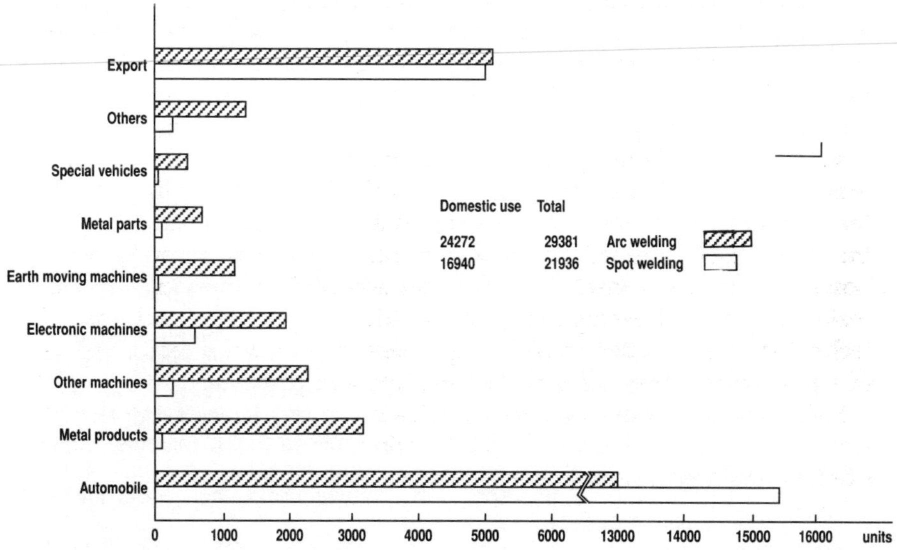


**Figure 12.2** Industry distribution of welding robots in Japan, 1988 (playback, NC and intelligent robots).

welding robots is concentrated in the automobile industry. Although a great number of arc-welding robots are used in the automobile industry, they are also used in various other industries.

## 12.1 SPOT-WELDING ROBOTS

Most spot-welding robots are found in the automobile industry, where they are used for white body spot-welding. Almost ten years have passed since automobile companies first began to use such robots. This

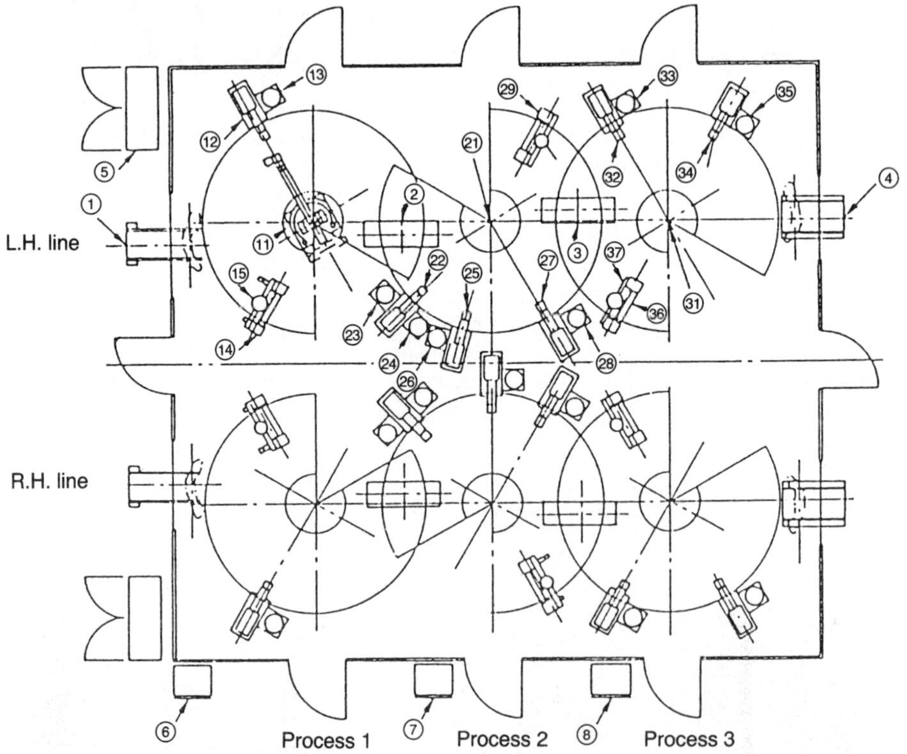


Figure 12.3 Spot-welding system for automobile parts: 1, workpiece loading unit; 2, 3, workpiece tables; 4, workpiece delivery unit; 5, LH overall control unit; 6–8, robot control units (YASNAC ERC) for processes 1–3. Process 1: 11, Motoman L60S robot; 12, fixed-position spot-welder; 13, nut feeder for M6; 14, material handling table; 15, material handler. Process 2: 21, Motoman L60S robot; 22, fixed-position spot-welder; 23, nut feeder for M10 round nut; 24, nut feeder for M10 oval nut; 25, fixed-position spot-welder; 26, nut feeder for M8 nut; 27, fixed-position spot-welder; 28, nut feeder for M10 square nut; 29, material handling table; 30, material handler. Process 3: 31, Motoman L60S robot; 32, fixed-position spot-welder; 33, nut feeder for M8 nut; 34; fixed-position spot-welder; 35, nut feeder for M6 nuts.

period has seen advances in robot utilization technology and improvements in robot performance and reliability. It has now reached a stage where spot-welding robots play an integral part in the automobile body-welding process.

The primary reasons for this shift to robots in the automobile-welding processes, which had previously been done manually, are:

(a) workers are relieved of work which is laborious due to the heavy spot-welding gun;
(b) the quality of work is improved since the robot does not miss any of the programmed weld points – a manual worker may miss some;
(c) robots provide greater flexibility by acting together with a system which can respond quickly to model changes. Thus, different types of automobile can easily be produced.

All of the robots described above are of the spot-welding type. Recently, robots which have been installed on sub-assembly lines have been fixed spot-welders. This type of robot picks up the workpiece brought to it on the line as shown in Figure 12.3. This system is a line in which the workpiece, in this case an automobile side frame, is held for nut projection welding, and consists of six robots.

Figure 12.4 shows a system consisting of a line in which automobile

Figure 12.4 Spot-welding robot holding spot gun.

side frames are conveyed and spot-welded automatically by robots. In order to reduce wear on the secondary cable between the transformer and spot-gun, the transformer is also held by the robot. Since transformer capacity has increased and automobile pre-coated steel sheets are no longer used, the payload of this robot has also been increased.

## 12.2 ARC-WELDING ROBOTS

Arc-welding robots were introduced in Japan around 1980, when subcontractors in the automobile industry began to use them in assembly lines. At that time, arc-welding of small parts was mainly performed by sub-contractors. There was also a shortage of new labour, and skilled labour was becoming scarce. The benefits of arc-welding robots were quickly recognized, resulting in their instant popularity. The major advantages of arc-welding robots can be summarized as:

(a) uniform product quality can be guaranteed;
(b) production management and productivity are improved;
(c) the dependence on skilled workers, who are always scarce in number, is reduced.

The system illustrated in Figure 12.5 is designed to improve a robot's operating efficiency by providing it with two interior jigs. These are used alternately for welding operations. Due to the fact that a worker must continually load and unload workpieces, the worker's workload increases considerably if the cycle times are shortened. In spite of this, the system has proven to be ideal for simple workpieces, and is still in operation today.

Previously, several arc-welding robots were used in a line in order to cut labour costs further. The development of various software, and

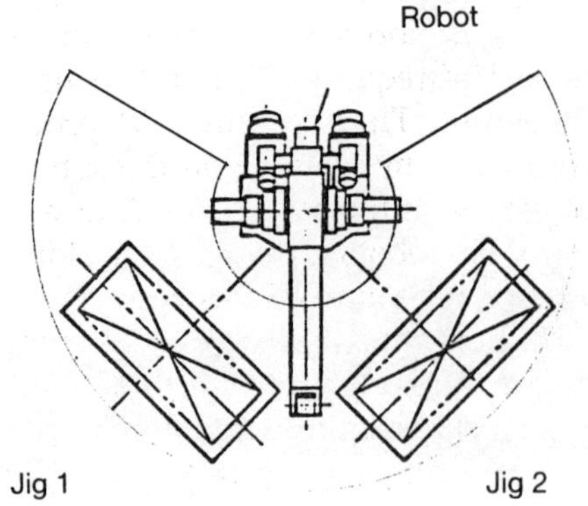


Figure 12.5 Arc-welding system.

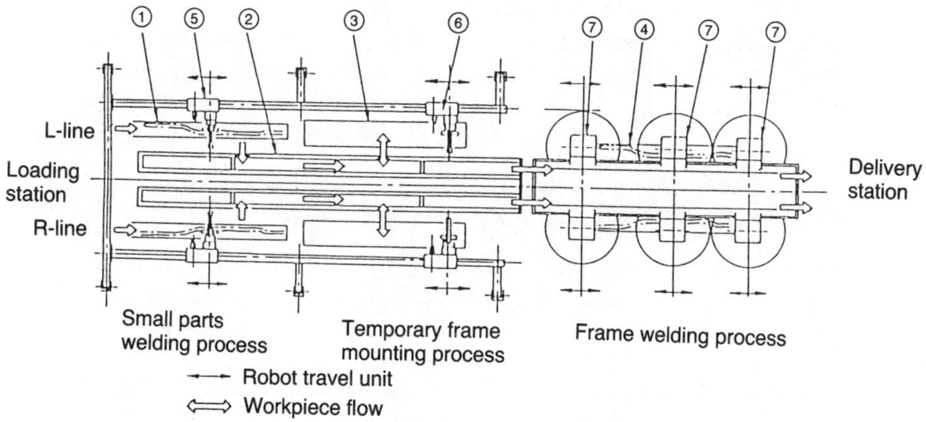


Figure 12.6 Arc-welding system: 1, small parts jig base; 2, conveying unit; 3, temporary mounting jig base; 4, welding jig base; 5, Motoman KGS; 6, Motoman KGS; 7, Motoman V5Z.

considerable improvements in the reliability and safety of robots, has made it possible to introduce lines containing tens of directly connected robots.

The arc-welding system shown in Figure 12.6 is a welding line where the workpiece, an automobile chassis frame, is conveyed automatically and welded by ten robots. This system has two lines. The workpiece from the loading station is set on the small parts jig base and welded by robots. The welded workpiece is then sent to the next process and set on a temporary mounting jig by the conveying unit. After temporary mounting, the workpiece is moved to the next robotic process and then conveyed to the delivery station.

The system illustrated in Figure 12.7 is chassis frame-welding agricultural equipment. The system has four jigs in each jig stocker (two large, two small). This configuration makes the system suitable for multiple small-lot production. The required jig is called up and the pieces set manually for arc-welding. Thus the finished product is obtained by assembling the workpieces by welding on the sub-assembly line.

Figure 12.8 illustrates an automatic arc-welding system for automobile chassis frames. Thirty-six robots are used for welding. Robot reliability is of prime importance on a line of this type. A mean time between failure of over 15 000 hours has been achieved with these robots. It is clear that the position of the arc-welding robot has changed from a skilled machine at a workstation to a robot on a production line. Robot technology alone is no longer sufficient to meet these demands. A complete integration of robot and production technologies is required.

The progress made so far in arc-welding technology should be noted.

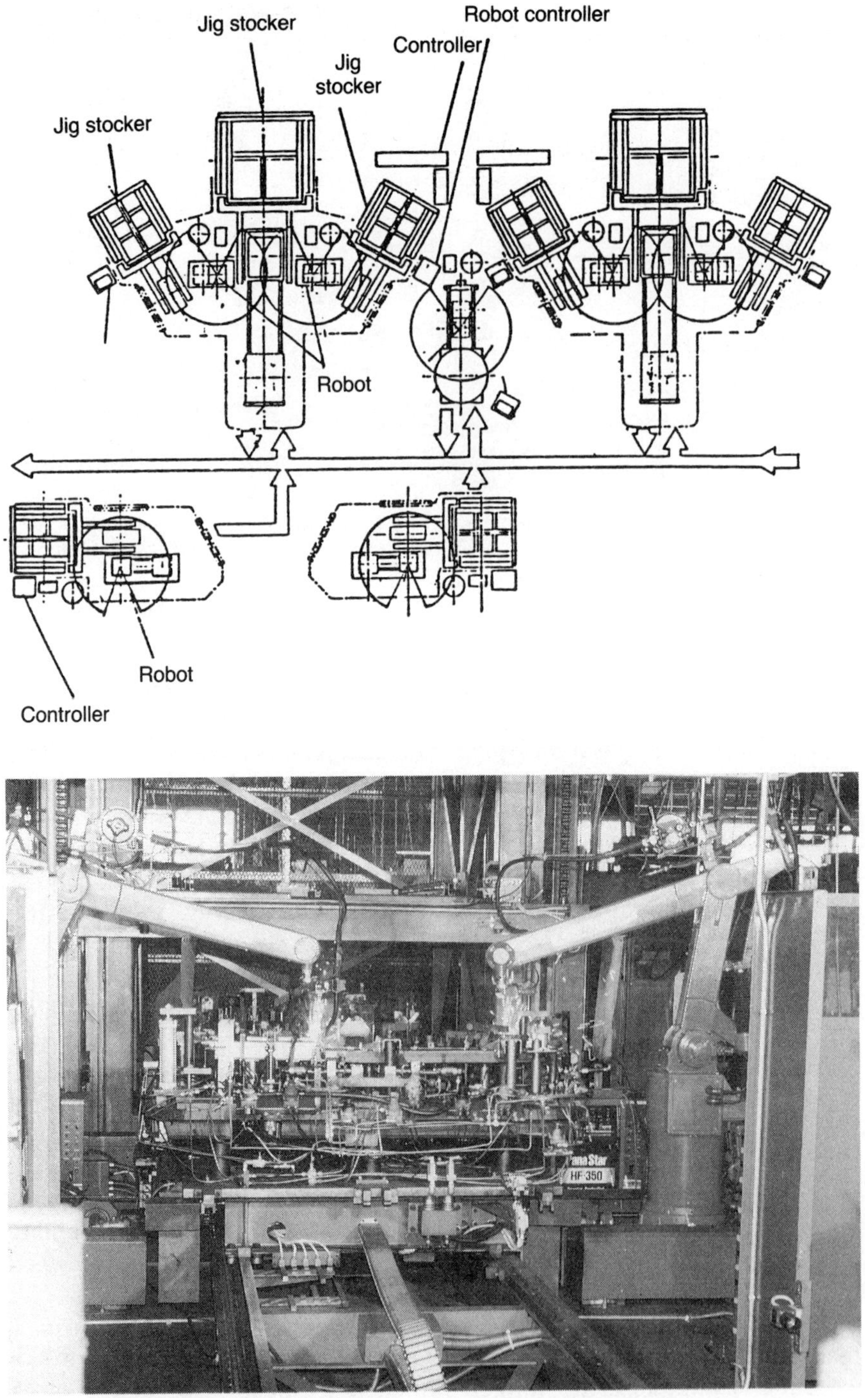


Figure 12.7 Seven-robot arc-welding system.

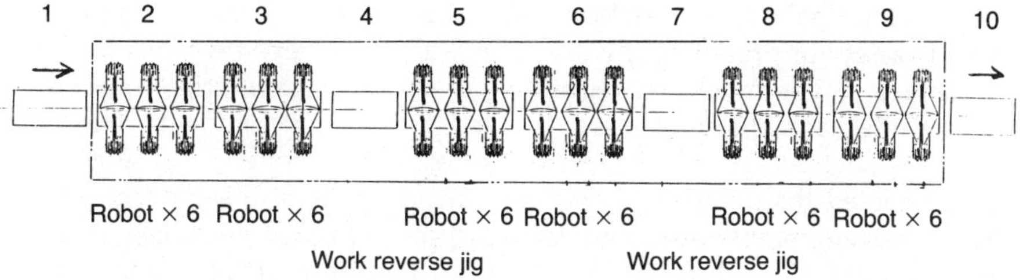


Figure 12.8 36-robot arc-welding system.

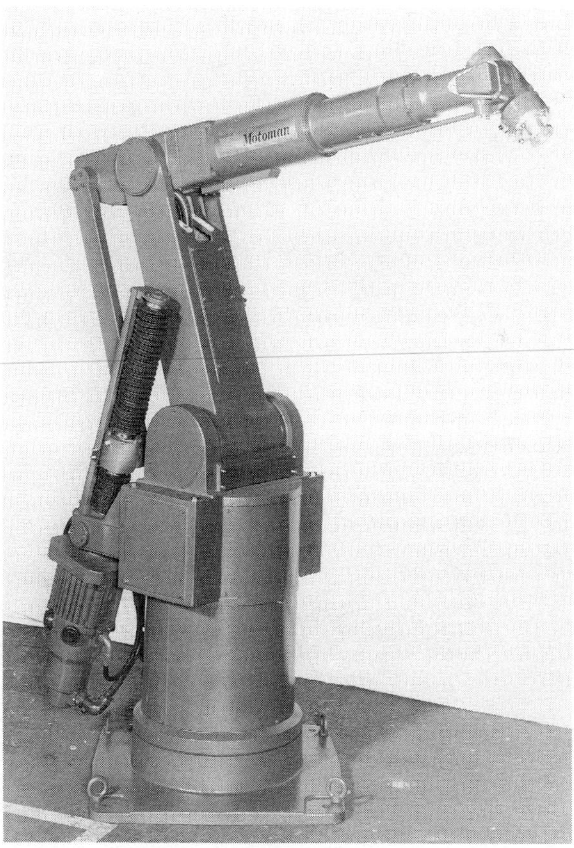


Figure 12.9 Standard arc-welding robot.

In conventional systems, which were geared to mass production of a small variety of products, the only aim was to improve the precision of workpieces. The increased requirement for small-lot multiple models, or complex workpiece shapes, has led to the development of technologies such as arc-sensors, vision sensors, and molten pool sensors. This enabled the detection of the arc starting point and seam tracking. These developments also lead to the implementation of robots in the manufacture of workpieces which do not require great precision. Arc-welding adaptive control, based on the molten pool sensor, is also reported to be nearing completion. Robots are gradually changing from the conventional five-axis type to the six-axis type to fully use the features of robot welding systems. In addition, functions such as off-line programming and computer communication are also developing.

A standard modern arc-welding robot is shown in Figure 12.9.

# Robot painting systems 13

*Toshihiko Nishimura*

## 13.1 CURRENT ROBOTIC PAINTING

Industrial robots have been used for painting for over 20 years. Kobe Steel Ltd (Japan) entered into a business relationship with Trallfa (Norway), a manufacturer of spray-painting robots. In 1973 they developed new hydraulic and electric-drive painting robots – Kobelco spray-painting robots.

Electric a.c. servomotors are now usually used in painting robots as they require less maintenance and provide higher reliability. They are smaller drive units than the hydraulic variety. It is, however, essential to prevent an electric painting robot from exploding due to ignition of flammable paint solvents.

It is important to be able to control the position, speed and orientation of a robot's spray gun. Data communications between robots and computers is crucial in a fully automated manufacturing cell.

Painting robot requirements, particularly for automobile manufacturers, are as follows:

(a) high path accuracy (within 5 mm) at high painting velocities (up to 1.2 m/s);
(b) multiple programming methods, such as teach pendant and off-line programming;
(c) line tracking to paint workpieces on moving conveyors;
(d) input/output interfaces to external manufacturing devices;
(e) computer communication capability for computer-integrated manufacture (CIM);
(f) safety, explosion proofing, self-diagnosis, and error detection.

*Manufacturing Automation Systems and CIM Factories*. Edited by K. Asai, S. Takashima and P.R. Edwards. Published in 1994 by Chapman & Hall, London. ISBN 0 412 48230 4

## 13.2 FEATURES

The Kobelco painting robot provides the features described below to meet these requirements.

### 13.2.1 HIGH SPEED AND ACCURATE MOTION

The maximum painting speed for a high-quality finish is approximately 1.2 m/s. The robot arm must be moved as quickly as possible between painting tasks to reduce the cycle times.

The manipulator arm can move the spray gun at up to 2 m/s with little vibration. The path, orientation and speed of the gun moving between points are calculated accurately by a microcomputer which controls the manipulator's servo axes in real time.

### 13.2.2 PROGRAMMING METHODS

Two programming methods are available with Kobelco robots, on-line programming and off-line programming.

On-line programming is that in which an operator moves the robot by remote control and records the current position and orientation of the robot arm using a teach pendant. Typical instructions such as velocity, input/output control, and control of the spray gun can be programmed with the pendant. A number of robot programs can be linked into one module, which is executed only once.

Off-line programming is that in which program development takes place on a personal computer, without using actual robots. Robots and work cells are animated on a graphics display. In these three-dimensional animations, the motions of robots and work cells can be simulated and verified. Completed programs are transferred from the computer to the robot controllers.

### 13.2.3 LINE TRACKING

External position detectors on a moving conveyor automatically adjust a robot's position, allowing a robot to paint workpieces on a conveyor.

### 13.2.4 EXTERNAL INPUT/OUTPUT INTERFACES

A robot, in a robotic painting cell, is likely to be required to control or synchronize external manufacturing devices. The robot treats several digital inputs as interlock signals, causing it to pause while a signal is not active. Several external digital input/output handlers are provided for fully automatic painting.

### 13.2.5 DATA COMMUNICATION

Several robots and computers can be connected to the Kobelco network, through which program data, control commands and status messages can be sent. The network system makes it easy to build up CIM systems.

### 13.2.6 SAFETY

The manipulator has a pressurized explosion-proof construction to make it safe in all painting environments.

When the power is turned on, the control unit automatically performs self-diagnostics and displays any problems which are found. Control circuits, the power supply, and operations are continuously checked for safety.

The specifications of the KRE410/420 Kobelco electric painting robots are given in Table 13.1.

**Table 13.1** KRE410/420 specifications

| Manipulator | KRE410 | KRE420 |
|---|---|---|
| Degree of Freedom | | |
| 1. Turning range (arm) | 300° | |
| 2. Forward/backward motion (arm) | 120° | |
| 3. Vertical motion (arm) | 120° | |
| 4. Rolling range (wrist) | 420° | |
| 5. Bending range (wrist) | 240° | |
| 6. Rotation range (wrist) | 420° | |
| Speed | Max. 2.0 m/s | Max. 1.5 m/s |
| Rated weight capacity | 8 kg | |
| Repeatability | ±1 mm | |
| Explosion protection | Combination of an air purged type and intrinsically safe type | |
| Ambient temperature range | 0~40°C | |
| Weight | 470 kg | 450 kg |
| Control unit (G type) | | |
| Memory unit (Maximum memory capacity) | | |
| Main Memory | IC Memory (with battery back up) | |
| IC Memory 128 kB (standard) | 3100 points | |
| IC Memory 768 kB (optional) | 19 400 points | |
| Number of programs | 998 | |
| Number of modules | 4000 | |
| Data-copying device | Floppy disk (optional) | |

**Table 13.1** *Continued*

| Manipulator | KRE410 | KRE420 |
|---|---|---|
| Operation<br>Teaching method | Remote teaching<br>Base coordinates (*x*,*y*,*z* operation)<br>Gun coordinates (*x*,*y*,*z* operation)<br>Joint coordinates (each axes) | |
| Recording method | PTP | |
| External I/O: input<br>Memorized signals<br>Command signals | <br>10<br>26 | |
| External I/O: output<br>Memorized signals<br>States signals | <br>17<br>16 | |
| Random program queuing function | Max. 100 programs | |
| Timer/counter function | 5 kinds | |
| Correction function | Position change, insertion, deletion<br>Changes of spray, output signals<br>Speed change between partial points<br>Change of input memorized signals<br>Module insertion, deletion | |
| Playback function<br>Playback check function | <br>Continuous, step, forward, reverse, normal speed, slow speed. | |
| Interpolation function | Linear, circular, smooth (optional) | |
| Speed control | Time (period) setting<br>Gun speed setting<br>Acceleration and deceleration | |
| Conveyor synchronization | Teach stationary, playback moving | |
| Changes in speed | Overall speed ratio setting ratio for each program | |
| Editing function | Program link, module editing | |
| Data control and display | Playback cycle, time, number of points.<br>Conveyor speed, shift distance.<br>Spray delay time, speed ratio, totalization of spray time, others. | |
| Self-diagnosis, error-detection function | CPU, memory, motor, robot operation | |
| Malfunction and operation error display | About 600 kinds | |
| Operation panel | Portable, with intrinsically safe type | |
| Ambient temperature range | 0–40°C | |
| Electric power supply | AC200V/220V, 3-phase, 50/60 Hz<br>7 kVA | |
| External dimensions, weight | 750(W) × 535(D) × 1600(H) mm,<br>260 kg | |

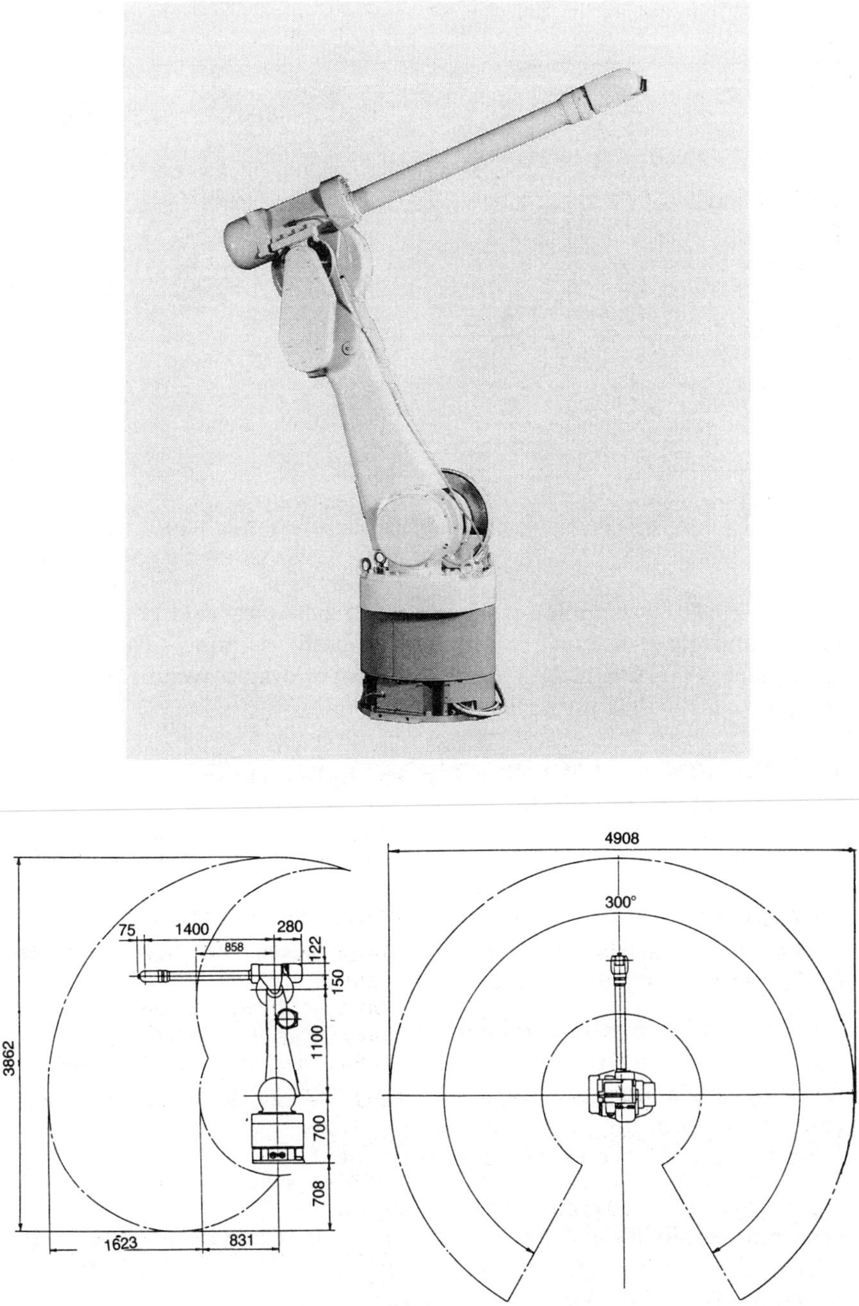


Figure 13.1 KRE 410 manipulator externals (dimensions in mm).

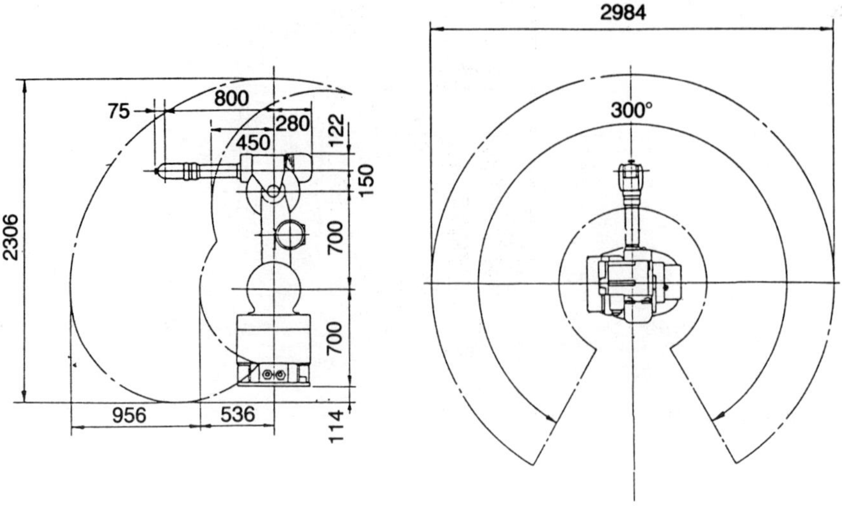


Figure 13.2 KRE 420 manipulator externals (dimensions in mm).

The KRE410 manipulator (Figure 13.1) occupies a large operating space, and its main use is automobile body painting. The KRE420 manipulator (Figure 13.2) is more compact and light-weight. It can be installed on the ceiling, wall or floor.

## 13.3 CONSIDERATIONS BEFORE INTRODUCING PAINTING ROBOTS

Successful robotic installation needs careful preparation and forethought. The design of a robotic cell requires expertise in robotic applications.

The following considerations should be taken into account when designing robotic painting cells:

(a) the painting machine to be used;
(b) how to transfer workpieces;
(c) process planning (e.g. pre-process, dry process);
(d) the layout of the robots and equipment;
(e) interfaces between robots and equipment.

After performing design work according to the specifications, the economic feasibility and payback period on investment must be determined. A system engineer from the robot manufacturer can assist in designing the robot system if required.

Kobe Steel Ltd has the expertise to design and implement robot painting systems for end users. Some typical applications are as follows

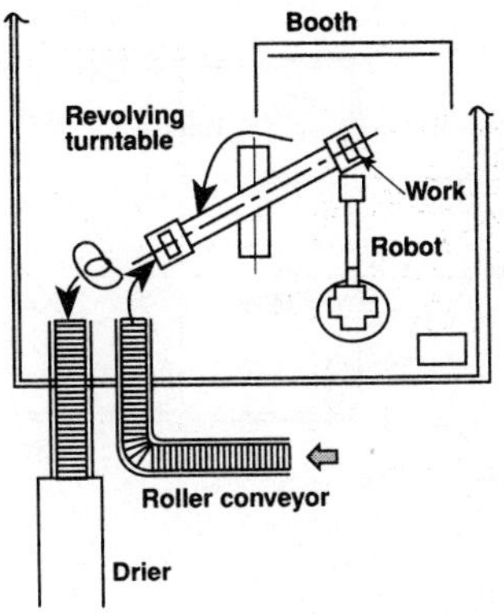


Figure 13.3 Painting domestic electrical appliance parts.

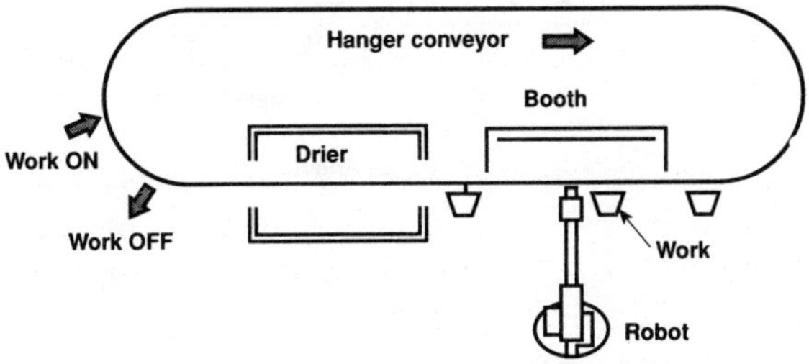


Figure 13.4 Painting woodwork.

(a) **Painting domestic electrical appliance parts**
   (i) Work – plastic frames for radio cassette recorders (Figure 13.3).
   (ii) Paint – melamine system paint.
   (iii) Effects – substitute for skilled painting operators; painting can be performed even by inexperienced operators, resulting in decreased labour costs; improvement of painting quality; reduction of the failure ratio; improvement in productivity by 50%.

(b) **Painting woodwork with legs**
   (i) Work – wooden chair (Figure 13.4).
   (ii) Paint – aminoalkyd resin, using electrostatic airgun.
   (iii) Conveyor – 2.4 to 2.8 m/min.
   (iv) Effects – number of workers reduced from four to one; painting costs reduced from ¥100/piece to ¥28/piece; failure ratio reduced from 3% to almost 0%; work changing time from one hour to almost zero time.

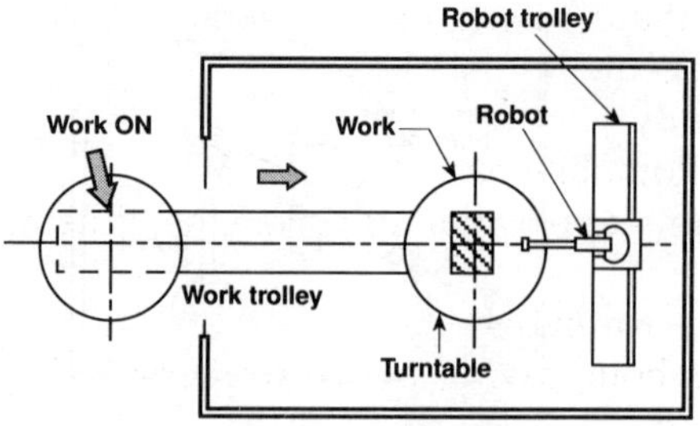


Figure 13.5 Engine-block cleaning.

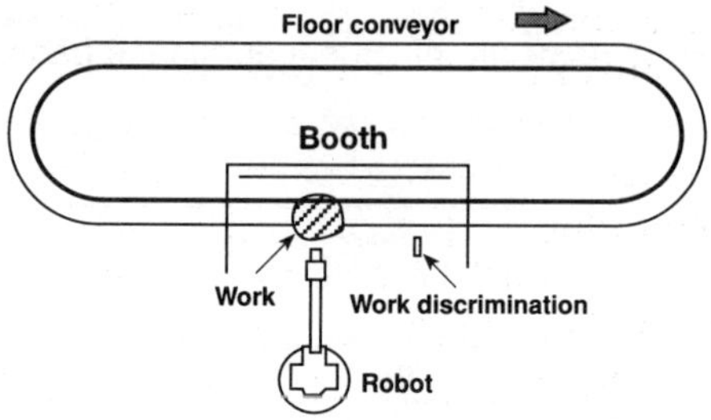


Figure 13.6 Application of mold-releasing agent.

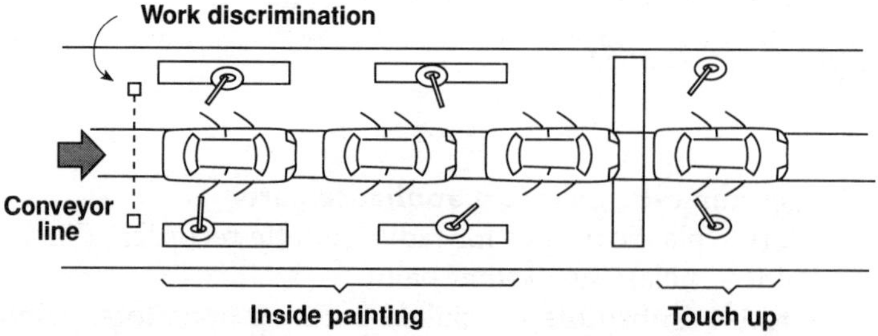


Figure 13.7 Application of surface solid to automobile body.

(c) **Engine block cleaning**
   (i) Work – cast iron diesel engine (Figure 13.5).
   (ii) Cleaning solution – water.
   (iii) Effects – labour saving of 1.5 workers.

(d) **Application of mould-releasing agent**
   (i) Work – automobile seat (Figure 13.6).
   (ii) Application – application of mould-releasing agent using an airless gun.
   (iii) Conveyor – 6–7 m/min.

(iv) Effects – labour saving of one worker, 30% saving in mould-releasing agent.

(e) **Automobile body**
(i) Work – automobile body (Figure 13.7).
(ii) Application – application of surface solid using an electrostatic gun.
(iii) Conveyor – 6 m/min.
(vi) Effects – labour saving of six workers; reduced failure ratio; paint saving.

## 13.4 FUTURE ROBOT PAINTING

In order to make a completely automatic painting cell, future robotic painting systems must provide new functions.

Modern robot sensor systems are being developed to take intricate measurements, such as the workpiece shape or position, and the thickness of a paint film. To perform more complex and flexible tasks, future painting robots must be able to use feedback signals or other information from intelligent sensor systems.

It is usual for multiple robots, such as spray-painting robots and door, bonnet or boot-opening robots, to work in the same painting cell. An integrated control system is required to control several robots and various items of equipment. New programming systems, such as expert systems, are required to implement a 'skilled painting robot'. These need to make decisions based on sensory data and internal knowledge.

# Index

Action control (of industrial robots) 67
Actuators, industrial robot 68
Adaptability index (FMS evaluation) 175, 176
Adaptively controlled robot 32, 80
Advanced Robot Technology Research Association (ARTRA) 27, 47
Advanced Robotics Project 80
  application fields 46
  elemental technologies covered 46
Advanced robots 80–1
Agriculture, use of industrial robots 31, 37
Andon device 160, 161
Appliance manufacture
  FMS used 173–9
  *see also* Automobile . . .; Domestic . . .; Electrical appliance . . .
Arc-welding robots 63
  adaptive control used 77, 240
  advantages 236
  compared with spot-welding robots 232
  number in Japan 40–1, 233
    distribution by industry type 233
Area (manufacturing floor) control systems, in CIM 107–8, 215
Area utilization index (FMS evaluation) 176
As-required approach (to CIM construction) 135
Assemblability evaluation method (AEM) 219
  assemblability improvements resulting 223
  cost estimation using 220
  design evaluation and improvement using 221
  design improvements achieved for VCR mechanism 222
  features 219
  indices used 220
Assembly-line performance
  improvements 228–30
  simulation 221, 223
Assembly robots 63–4
  compliance considerations 126–7
  control motor manufacture 186, 187, 189–92
  future developments 230
  intelligent adaptive control systems used 78, 79
  number in Japan 40–1
  VCR mechanisms manufacture 218–30
  with video cameras 189–92
Australian Robot Association 53
Austria
  industrial robots
    firms involved 45
    numbers in use 33
  robot/automation association 55
Automatic assembly 61–2
  motor parts 186–7
  VCR mechanisms 224–5
    construction of assembly stations 225–8
Automatic collection of data 109, 123
Automatic guided vehicles (AGV) 65–6, 225
Automation, *see* Factory automation

Automobile industry
CAD/CAM systems used
background 195–8
future developments 206–9
implementation 198–206
industrial robots used 38–9, 233, 234–40
painting robots used 248, 249
product development systems 194–5
style design systems 207–8
welding robots used 233, 234–40
Autonomous distributed/ decentralized control-integrated manufacturing system (ADCIMS) 85, 86

Belgium
industrial robots, numbers in use 33
machine tool imports from Japan 21
robot/automation associations 53, 54
Bottom-up approaches
business growth 105, 112
CIM construction 135
Broadband communication cable/channels 157, 158
Burn-in method of reliability testing 132

Canada, industrial robots, numbers in use 33
Capital investment, increase in return by using industrial robots 51
Cartesian coordinate robots 81, 177
Centralization of authority, integration affected by 7–8
China, robot/automation association 55
Civil/construction engineering, use of industrial robots 28, 31, 37
Communication control (of industrial robots) 67–8, 241
Competitiveness 100
and CIM systems 106, 213
key objectives required 103
Compliance-controlled soft motion robot 126
COMPOSER CNC system 81
CAD component 82
CAM component 82
communications components 81–2
concept of 82–3
evaluation 83–4
Computer-aided design (CAD)
applications, automobile industry 194–209
benefits 201, 205–6
with CIM/automation systems 107, 119–20, 200, 203, 215
in COMPOSER system 82
definition 119–20
functional elements 120
reasons for use 120
Computer-aided inspection (CAI) 129
Computer-aided manufacturing (CAM)
applications, automobile industry 194–209
with CIM/automation systems 120–1, 200, 203
in COMPOSER system 82
definition 120
Computer-aided quality control (CAQC) 128–33, 161–2
Computer-aided stress testing (CAST) 132
Computer-aided test (CAT) 129–30
Computer industry, CIM used 210–17
Computer-integrated assembly systems 125, 126
Computer-integrated manufacture (CIM) 105–10
applications
computer industry 210–17
machinery industry 155–65
benefits 216–17
CIM in series 6, 7

Computer-integrated manufacture *contd*
communication standards for 90–5
construction of system 133–5
as-required approach 135
bottom-up approach 135
considerations affecting 133–4
prioritized-step approach 134–5
project steps involved 135
top-down approach 134–5
cross CIM 6, 7
factors affecting 88
future developments 60, 135–6
goals 118
implementation of 110–18
basic steps 109–10
bottom-up approach 105, 112
comprehensive plan 113–14
continuous improvement effects 114–16
education/training programmes in support 117–18
enhancement effects 114, 115
by IBM 111
problems encountered 136
project organization 111–12
step-by-step implementation 114
systems engineering used 8–12
top-down approach 105, 112, 113
meaning of term 57
objectives 105–6, 131, 155
practical manufacturing systems 89–90
progress towards 4, 5, 58–60, 61, 101–3, 214
reasons for use 4, 6–8, 118–19
safety aspects 137–52
system hierarchy 107–8, 214–16
administration layer 107, 108, 214, 215
engineering information systems 107, 108, 214–15
management layer 107, 108, 214, 215
materials logistics systems 107, 108, 215
production control layer 107, 108, 215
system integration 106–7, 216
system security 118
systems in Japan 88–9
technologies involved 57, 108–9
types 6
use in factory automation 85–90, 118–28
*see also* Factory automation
Computer-interpreted test 129, 130
Computer numerically controlled (CNC) robots, number in Japan 38–41
Computer numerically controlled (CNC) system 81–5
communications components 81–2
Continuous flow manufacturing (CFM) 103
integration with CIM/ automation 104, 105
Control motor manufacture
factory automation for 180–93
stator-wire lead extraction 189
automation of 189–90
tie-bolt nut tightening 190–1
automation of 190–2
Cost control system 161
Customized products 102, 103, 121, 167

Deburring machines 84
Deburring robots 64–5, 84–5
intelligent adaptive control systems 77–8
Denmark
industrial robots, numbers in use 33
robot association 53
Domestic electrical appliance parts, painting of 247

Economic cycles
  effect on Japanese labour force 34
  machine tool industry affected by 15–16
Electrical appliance parts, painting of 247
Electric discharge machines (EDMs)
  orders situation in Japan 18
  production statistics for Japan 19
Employment opportunities, effect of industrial robots 50–1
Engine-block cleaning 248
Engineering support systems (in CIM) 107, 108, 215
Europe
  industrial robots
    international industrial cooperation 45
    numbers in use 33
  machine tool imports/exports (to/from Japan) 20, 21, 22
  robot/automation associations 52, 53, 54, 55
Exports (from Japan)
  industrial robots 37
  machine tools 19–20, 21
Extreme-environment robotics, development of 26–7, 47

Factory automation (FA)
  CIM systems used 85–90, 118–28
  computers used 121–3
    functions 122
  configuration 162–4
  control motor manufacturing example 180–93
    assembly operations 186–7
    economic efficiency of automated factory 192
    factory layout 181–4
    inspection operations 187
    machining operations 184–6
    technological developments used 187–92
  cost control system 161
  effect of socio-economic factors 5
  equipment maintenance control system 162
  example of FA system 7
  experimental training system 162–5
    configuration 162–5
  fifth phase 59–60
  first phase 58
  fourth phase 59
  in Japan 12–25
  meaning of term 4
  numerical control engineering used 81–5
  production control system 161
  quality control system 161–2
  safety aspects 142
  second phase 58
  significance 4, 6–8, 25
  systems engineering for introduction 8–12
    advantages/disadvantages predicted 10
    equipment selection 11
    examination of existing equipment 9
    examination/improvement of current manufacturing processes 10
    in-house announcement 10
    new system design 11
    operation selection 10
    planning and execution stage 12
    post-introduction engineering 12
    review of available automation methods 9–10
    staff planning/recruitment considerations 11
  third phase 58–9
  *see also* Computer-integrated manufacture (CIM)

Fail-safe AND gate 146–8
analysis of faults 148
circuit diagram 147
input/output relationships 148
Fail-safe technology 137, 138, 143–5, 152
example of use 144–5
*see also* Safety confirmation system
Fanuc control motor factory 180–93
automatic assembly operations 186–7
automatic inspection operations 187
automatic machining operations 184–6
economic efficiency 180, 192
layout 181–4
parts flowchart 183
technologies involved 187–92
FAROT-M6 robot 78
Finishing (of machine parts) 84–5
*see also* Deburring machines; Deburring robots
Finland
industrial robots
firms involved 45
numbers in use 33
Robotics Society 54
Fixed-sequence control robots
definition 32
production in Japan 31
Flexible manufacturing cells (FMC)
application for rotating parts production 169
growth in use in Japan 23
Flexible manufacturing systems (FMS)
advantages 166, 167
aims/objectives 6, 174
application examples
appliance manufacture 173–9
machine tool industry 167–71
rotating parts 168–70
sheet metal parts 170–1
small cubic parts 167–8
case studies 87
equipment/robots used 132, 178
evaluation measures 174–5, 176
first introduced 4
history 102, 103
number in Japan 23, 24
prior adjustment of production conditions 175, 177
progress towards realization 175, 177
system configuration 23
modular approach 178
Forestry, use of industrial robots 31, 37
France
industrial robots
firms involved 45
numbers in use 33
robot association 53
Fujisawa plant (IBM) 212–13
CIM used 214–16

Gas hot-water heater, fail-safe technology used 144–5
General-purpose simulation system (GPSS) 221
input and output 223
simulation results 221, 223
compared with actual line performance 224
Germany
industrial robots
firms involved 45
numbers in use 33
machine tool imports/exports (to/from Japan) 21, 22
robot/automation associations 53, 55
GNC2 CAD/CAM system 196
applications 198, 200, 201, 202–3
basic functions available 198
basic shapes used 197
benefits 205
operating equipment used 198, 199
promotion in Mazda 201

Hazard detection system 139
compared with safety confirmation system 139–41

Hitachi A3020 assembly robot 227, 228
Hold-stop state (for robot) 151
safety-control devices used 150–2
Holland, *see* Netherlands
Hop/skip/jump approach (to implementation of CIM) 114
Human–machine interfaces 87, 158, 160
Hungarian Robotics Association 54

IBM
and CIM 111, 213–16
Fujisawa plant 212–13
CIM used 214–16
global systems 210–12
robot system 126–7
Imports, machine tool, to Japan 20, 21, 22–3
Independence, integration affected by 7–8
Industrial Research Institute
FA training system 162–5
configuration 162–5
Industrial robots
advanced robots R&D 46, 80–1
advantages 25–6, 101
architecture 66–8
action control 67
communication control 67–8
motion control 66–7
in assembly lines 124–7, 186, 187, 218–30
design procedure 124–5
classification of types 32, 80–1
communications systems used 70–3
real communications systems 71, 72
virtual communications systems 71, 72
compared with single-task automatic machines 30, 32, 49, 51
configuration 62
definitions 29–30, 32, 124
drive mechanisms used 68–9
functions 62, 66
future developments 230
hold-stop state, safety-control devices used 150–2
international cooperation 42–7
exchange of information 42–3
industrial cooperation 43–6
research cooperation 46–7
in Japan 25–52, 61–81
distribution by industry type 36, 38–9, 233
distribution by process type 37, 40–1
factors affecting rapid growth 33–5
long-term demand forecast 28, 31
numbers in use 27, 28, 33, 37, 38–41, 101, 132, 233
policies to encourage 26, 27
production statistics 27, 29, 31, 101
standards 32, 78–9
languages used 73–5
manufacturing applications 35, 36, 37, 62–6
assembly systems 63–4, 186, 187, 218–30
deburring systems 64–5, 84–5, 188
MAGVR 65–6
painting systems 241–9
welding systems 63, 232–40
materials used 69
non-manufacturing applications 28, 31, 35
prospective roles 52
reliability 127
safety aspects 127–8, 148–52
see-saw mode operation, safety-control devices used 148–50
selection of 124
socioeconomic impact 47–52
capital investment advantages 51

Industrial robots *contd*
employment effects 50–1
humanization of working life 50
new-industry development 51
productivity improvement 49–50
working environment affected 50
*see also* Assembly ...; Intelligent ...; Numerically controlled ...; Operation ...; Painting ...; Playback ...; Sequence control ...; Welding robots
Information transmission device 160, 161
Inspection robots 164, 187
number in Japan 40–1
Integrated (information) network 116, 117
Integration, effect of centralized authority 7–8
Intelligent adaptive control robot systems 77–8
Intelligent management and manufacturing (IMM) system 86
Intelligent robots 32, 80
number in Japan 38–41
production in Japan 31
Interlocks (safety)
application to automatic machines 143, 144
fundamentals 142–3, 152
International Conference on Advanced Robotics (ICAR) 43, 56
International cooperation, industrial robots 42–7
International Federation of Robotics (IFR) 42, 54
members listed 52–5
International (information) network 116
International Organization for Standardization (ISO)
communications protocols/software 156
definitions/standards for robotics 66, 96–7
Japanese adaptations 89, 97–8
working groups on robotics 78–9
International Robotics and Factory Automation (IRoFA) Centre (Japan) 27, 47, 54
International Symposium on Industrial Robotics (ISIR) 42, 55–6
Investment effectiveness index (FMS evaluation) 176
Islands of automation 91, 129
*see also* Stand-alone systems
Italy
industrial robots
firms involved 45
numbers in use 33
machine tool imports from Japan 21
robot society 52

Japan
CIM systems 88–9
factory automation 12–25
IBM plants 212–13
industrial robots 25–52, 61–81
factors affecting rapid growth in use 33–5
international industrial cooperation 45
long-term demand forecast 28, 31
numbers in use 27, 28, 33, 38–41, 101, 132, 233
distribution by industry type 36, 38–9, 233
distribution by process type 37, 40–1
policies to encourage use 26, 27

production statistics 27, 29, 31, 101
standards 32, 78–9
labour shortages 34–5
painting robots 241
population growth 34
robot associations/societies 43, 46, 52, 54
welding robots 233
Japanese industrial Robot Association (JIRA) 52
international industrial cooperation 46
international seminars/ conferences 43, 44, 56
standards for industrial robots 89
Japanese Industrial Standards (JIS)
classification/definitions of industrial robots 32, 89
definition of machine tools 12–13
Japanese machine tool industry
characteristics 13–16
import/export statistics 19–23
orders situation 16–18
production statistics 13, 14, 15, 18–19
supply-and-demand position 16–23

Kanban (stock control) systems 157, 161, 216
Kobelco painting robots 241
applications 247–9
external dimensions 245–6
programming methods available 242
safety aspects 243
specifications 243–4
Korea, machine tool imports/ exports (to/from Japan) 21

Labour force, Japanese 34
shortage affecting use of industrial robots 34–5, 50, 236
Labour safety, industrial robots 127–8
Laser machining 171
Learning-controlled robot 32, 80
Limit switch (safety switch) 139
Local area networks (LANs) 86, 91, 116, 123

Machine tool industry
factors influencing future 24–5
FMS used 166–72
Machine tools
definition 12–13
supply and demand in Japan 16–23
import/export statistics 19–23
orders 16–18
production statistics 13, 14, 15, 18–19
Machining centres (MCs) 15
Japanese production statistics 14, 15, 18–19
Machining robots
motor parts 184–6
number in Japan 40–1
Maintenance
control system for 162
industrial robots 141, 143
Manipulating industrial robot (MIR)
definition 66
drive mechanisms used 68–9
functions 66
joint motions 68
Manufacturing automatically guided vehicle robot (MAGVR) 65–6
Manufacturing automation protocol (MAP) 85
characteristic features 156–7
communication standards 91, 156
conformance testing 92
documentation necessary 93
flow diagram 94
international harmonization 92

Manufacturing automation protocol (MAP) *contd*
test reports issued 93
test system 92, 93
practical realization 157–62
configuration of network 157–8, 159
in FA training system 162–5
man–machine interfacing 158, 160
test centre 91–5
Manufacturing technology, past and future developments listed 136
Market-in approach to manufacturing 103, 174
Materials logistics systems (in CIM) 107, 108, 215
Materials requirements planning (MRP) 107, 215, 216
Matsushita Group 174, 179
Mature engineers/operators
CNC system used by 81–4
*see also* Older people, employment of
Mazda's CAD/CAM system 194–209
applications 198, 200, 201, 202–3
basic functions available 198
basic shapes available 197
benefits 201, 204–6
development of 195–6
equipment used 198, 199
implementation 198–206
philosophy behind 196
Mechatronic machine tools, first developed 5, 58
Mechatronics Investment Promotion Tax system 26
Medical care/welfare, use of industrial robots 28, 31, 37
MeiRobo GWR60 robot 77–8
Methodology for Unmanned Manufacturing (MUM) 87
Mining, use of industrial robots 31, 37
Mollins System 24, 166
Motion control (of industrial robots) 66–7
Motoman robots 77, 232, 234, 237
Mould-releasing agent, application by robot 248–9

Netherlands
industrial robots
firms involved 45
numbers in use 33
robot/automation associations 54
Network integration 116, 117, 209, 241
Norway
industrial robots, numbers in use 33
robot/automation association 55
Nuclear power industry, use of industrial robots 28, 31, 37, 51
Numerical control engineering 81–5
Numerical control (NC) machine tools
CAD/CAM input 200, 203, 209
cutter interference calculations 205–6
first developed 5, 166
Japanese production statistics 14, 15, 18–19
Numerically controlled robots 32, 80
number in Japan 38–41
production in Japan 31

Ocean exploration/development, use of industrial robots 28, 31, 37, 51
Office automation (OA) 108, 123, 216
Older people, employment of
CNC system used 81–4
industrial robots used 50
Open systems interconnection (OSI) protocol 91
Operating robots 32, 80
production in Japan 31
*see also* Teleoperating robots
Operation improvement, CIM used 103–5

Painting robots 241–9
accuracy of motion 242
applications 247–9
data communications systems used 243
design considerations 246
external input/output interfaces used 242
features 242–6
future developments 249
number in Japan 40–1
programming methods used 242
requirements 241
safety aspects 243
speed of motion 242
Photoelectric safety switches 139–41
Physicochemical process machines
exports from Japan 20
imports to Japan 22
*see also* Electric discharge machines; Laser machining
Playback robots 32, 80
number in Japan 38–41
production in Japan 31
Poland, robot/automation association 55
Post-introduction engineering 12
Prioritized-step approach (to CIM construction) 134–5
Product development
automobile industry 194–5
VCR mechanisms 218–21
Production control systems
in CIM 107, 108, 161, 215
in FMS 177
Productivity 100–1
CAD/CAM systems affecting 206
CIM affecting 104, 105
definition 100
factors influencing 100, 101
factory automation affecting 192
industrial robots affecting 49–50, 247, 248, 249
Product liability (PL) 137–8
amount paid in compensation in USA 138
Product-out approach to manufacturing 103, 174
Prototyping, automobile industry 205, 207, 208

Quality assurance 114
Quality circles 112, 115–6
Quality control (QC) 128
computer-aided quality control 128–33, 161–2

Reliability
industrial robots 127, 229, 237
products
improvement of 130–1
tests 132–3
Robot associations/societies 52–5
Robot intelligence 75–8
Robot languages 73–5
classification 73–4
Bonner's classification 73
Inoue's classification 73
decision-making processes 74–5
primitive level languages used in Japan 74
Robotic painting 241–9
*see also* Painting robots
Robotics
development of extreme-environment robotics 26–7, 47
international research cooperation 46–7
Robots, *see* Assembly . . .; Industrial . . .; Intelligent. . .; Numerically controlled . . .; Operating . . .; Painting . . .; Playback . . .; Sequence control . . .; Welding robots
Rotating parts, FMS for 168–70

Safety aspects
CIM factories 137–52
industrial robots 127–8, 148–52
painting robots 243

Safety confirmation system
  basic concept 141–5, 152
  devices used 139, 140
  use in automated factory line 145–52
  *see also* Fail-safe technology
Safety control system 138–41
  applications
    hold-stop state 150–2
    see-saw robot operation 148–50
Sealed beam scanner 145–6
  advantages 146
Security control, CIM systems 118
See-saw mode robot operation, application of safety-control devices 148–50
Selective compliance assembly robot arm (SCARA) type robot 64
  languages used 73, 74
Sensory-controlled robot 32, 80
Sequence control robots 32, 80
  number in Japan 38–41
  production in Japan 31
Servo motors, assembly procedures 184, 186–7
Shape memory effect (SME)
  compared with servo control system 70
  use in robots 69
Sheet-metal parts, FMS for 170–1
Singapore
  machine tool imports/exports (to/from Japan) 21
  robot/automation associations 53
Socioeconomic factors
  automation progress affected by 5
  effect of industrial robots 47–52
Soft motion robot (SMR) 126
Space elasticity, industrial robots 49
Spain
  industrial robots
    firms involved 45
    numbers in use 33
  robotics association 54
Spot-welding robots 234–6
  advantages 235
  automobile manufacture using 234–6
  compared with arc-welding robots 232
  first developed 232
  number in Japan 40–1, 233
    distribution by industry type 233
  payload carried 232, 236
  servo-motor drives used 232
Stand-alone systems 91, 129
Statistic quality control, applications 128
Stress testing of components/products 132
Sweden
  industrial robots
    firms involved 45
    numbers in use 33
  robot association 53
Switzerland
  industrial robots
    firms involved 45
    numbers in use 33
  machine tool imports/exports (to/from Japan) 21, 22
  robot/automation association 55
Systems engineering (SE)
  and automation/CIM 8–12
  use with industrial robots 43

Taiwan, machine tool imports/exports (to/from Japan) 21
Tape recorder mechanisms, *see* Videocassette recorder (VCR) mechanisms
Tax incentives
  high-tech R&D 27
  mechatronics investment 26
Technical information management (in CIM) 109
Technology seminars, on industrial robots 43, 44
Teleoperating robots
  definition 32

Teleoperating robots *contd*
number in Japan 38–41
production in Japan 31
Thailand, machine tool imports from Japan 21
Time elasticity, industrial robots 49
Top-down approaches
business growth 105, 112, 113
CIM construction 134
Toyota Machine Works, MAP used 156, 157
Transportation industry, use of industrial robots 28, 31, 37

UK
industrial robots
firms involved 45
numbers in use 33
machine tool imports from Japan 21
robot association 52
Ultrasonic fail-safe sensor 140
Unimate 232
USA
firms involved in robotics 45
industrial robots
international industrial cooperation 45
numbers in use 33
machine tool imports/exports (to/from Japan) 20, 21, 22
product-liability compensation payments 138
robot associations/societies 52, 53, 54

Variable-sequence control robots
definition 32
production in Japan 31
Versailles (France) summit meeting, proposal on robotics 42
Video cameras
assembly robots using 164, 187, 189–92
inspection robots using 164
Videocassette recorder (VCR) mechanisms 219
automatic assembly for 224–5
benefits resulting 228–9
construction of assembly stations 225–8
design improvements 222
assemblability evaluation 223

Welding robots 63, 232–40
advantages 235, 236
number in Japan 40–1
*see also* Arc-welding robots; Spot-welding robots
Wide area network (WAN) 91
Williamsburg (USA) summit meeting, on international cooperation on robotics 42, 43, 46
Woodwork, painting of 247
Working environment, effect of industrial robots 50